The Patrick Moore Practical Astronomy Series

More information about this series at http://www.springer.com/series/3192

Viewing and Imaging the Solar System

A Guide for Amateur Astronomers

Jane Clark

 Springer

Jane Clark
Director of Observing
Bristol Astronomical Society
Long Ashton
Bristol BS41 9BQ
United Kingdom

ISSN 1431-9756   ISSN 2197-6562 (electronic)
ISBN 978-1-4614-5178-5   ISBN 978-1-4614-5179-2 (eBook)
DOI 10.1007/978-1-4614-5179-2
Springer New York Heidelberg Dordrecht London

Library of Congress Control Number: 2014946074

Printed on acid-free paper

Springer is part of Springer Science+Business Media (www.springer.com)

Preface

This book is written because it has become clear to me that many members of astronomical clubs and societies have a vague idea how to observe the Solar System, but not much detailed knowledge. They have often never used a webcam for example, or if they have, they had a bad experience because it did not produce brilliant results first time.

There seemed to be room for a 'how to' book that did not attempt to take you to being a professional or world class amateur, but rather would give you enough information to encourage you to get going. Very importantly, it should also reassure you that it is OK to produce lousy photos at first. I did not learn this stuff overnight, and nor will you. You will learn it over quite a lot of nights. It does take a bit of determination to produce good results.

I have also completely omitted an important method of Solar System observation: drawing and sketching. This is because I have no skill in that field. I have the greatest respect and admiration for those who can do it, but it is not my thing. I have been a photographer for 40 years and regard it as my imaging medium, although I freely admit that drawing a scene on the Moon does make you think about what's there more than photography does. I became good enough as a teenager to win a handful of prizes for terrestrial photographs of a completely non-scientific nature. The pictorialist in me still occasionally surfaces, for example in Fig. 7.6.

There is something of a philosophical issue about using software to enhance astronomical images: what is 'really' there? Is it really more honest than making a drawing, where the artist has to decide what to include and exclude? I think the answer is, 'No it is not.' Astrophotography is a subjective medium.

My policy on referencing work has been roughly as follows. I have tried to give at least secondary sources, such as books, for other people's ideas, measurements and discoveries. This is much easier in fields where I do not participate, such as providing general information about the planets, because I did not learn any of this stuff by any means other than reading. It has been much more difficult to cite sources for the chapters on equipment and technique, because I learned only some of this stuff by reading. I gathered a lot of the information by watching what fellow astronomers were up to, and by trying out different telescopes etc. I suppose I am relatively gung-ho about trying a technique, and being patient and persistent when it all goes horribly wrong the first time. Therefore these chapters are quite light on citations.

Although astronomy is not an instant gratification game, over the last few years I have gained a lot of satisfaction from my astronomy. It has been worth the hassle, the frustration, the need for patience, the lost sleep and the getting cold. If I can infect you with some of my enthusiasm, I will have succeeded.

Bristol, UK   Jane Clark

Acknowledgements

I would like to thank members of the various astronomical societies to which I have belonged, Norfolk and Norwich, West Norfolk, the Society for Popular Astronomy and now Bristol, for encouragement and advice over the years. In particular, although they have probably forgotten what they taught me, I would mention Freddy Rice and Adrian King for introducing me to many techniques, Sue Napper for introducing me to the idea of webcams, Darren Sprunt for much detailed advice, and Trevor Nurse for showing me about binoculars, and Robin Scagell for sending me some advice via the Society for Popular Astronomy's bulletin board on how to process images of Jupiter with Registax. The small hints these people fed to a receptive enthusiast went a long way.

I would also like to thank the staff of Springer, notably Maury Solomon and John Watson, for encouragement with this book project. My personal life went through a very rough patch recently, and writing this book has been good therapy to help me rebuild my confidence.

Contents

About the Author

Jane Clark is an English amateur astronomer who earns her living as an engineer. She has a Ph.D. in physics and an MBA from Warwick University. She completed 2 years of postdoctoral training at Case Western Reserve University in Ohio before returning to England to begin an industrial career. She became interested in both astronomy and photography as a teenager in the 1970s, photography much more seriously, although as her career progressed and family commitments increased, both interests lapsed. She acquired a telescope in 2006, shortly after completing her MBA, and quickly became hooked on observing. This experience made her realize that astronomy is a lot more fun than business administration. She is a member of Bristol Astronomical Society, and was a founder member of West Norfolk Astronomy Society. Jane gives talks on the Solar System to astronomy clubs, and other societies as diverse as the cub scouts and church wives' groups, and helps with the public outreach activities of her club in Bristol.

Chapter 1

How to Find the Solar System

The Solar System: An Obvious Concept?

Even a book for non-astronomers would not need to explain what the Solar System is. Everyone knows.

Well, almost everyone. Plus, we are still discovering the outer reaches of the Solar System. With those caveats, the principal members of the Solar System are as follows.

Analysis of Table 1.1 shows that it is meaningful to divide the Solar System up into regions. The rocky planets orbit within 2 AU of the Sun. There is then a much larger region from 5 to 30 AU occupied by the gas giants. People talk about an inner Solar System, containing rocky planets, and an outer Solar System, containing gas giants, conveniently divided by the Asteroid Belt. With the ongoing discovery of the inhabitants of the Kuiper Belt, this terminology may become obsolete. The gas giants may turn out to inhabit the middle Solar System.

We did not always know this information. The Solar System had to be discovered. It was not discovered overnight. Indeed dwarf planets further away than Neptune (TNOs, or trans-Neptunian objects) are still being discovered [14]. The alleged Oort Cloud, many times as far from the Sun as Neptune, has yet to be detected with any certainty. If you find it, apply to the King of Sweden for a Nobel Prize.

Much if not most of what we know about the planets has been discovered using space probes. As a result we live in a golden age of Solar System discovery, the like of which has not been seen since the heady days of Copernicus, Kepler,

© Springer Science+Business Media New York 2015
J. Clark, *Viewing and Imaging the Solar System*, The Patrick Moore
Practical Astronomy Series, DOI 10.1007/978-1-4614-5179-2_1

Table 1.1 Principal members of the Solar System [1]

Name	Type of object	Distance from Sun (AU)	Diameter (compared to Earth)	Mass (compared to Earth)	Brightness (magnitude)[a]	Years to Orbit Sun
Sun [2]	Star	–	109	333,000	–27	–
Mercury	Rocky planet	0.31–0.46	0.38	0.05	+5.7 to –2.6 [3]	0.24
Venus	Rocky planet	0.72	0.95	0.81	–3.8 to –4.9 [4]	0.62
Earth	Rocky planet	1.00	1.00	1.00	–	1.00
Earth's Moon	Satellite	1.00	0.27	0.01	–2.4 to –12.9 [5]	1.00
Mars	Rocky planet	1.52–1.66	0.53	0.11	+1.6 to –3.0 [6]	1.88
Asteroid Belt [7]	Belt of rocks	1.5–5 approx	0.04 to dust speck	0.0002 to dust speck	+∞ to +6.7	2–10
Jupiter	Gas giant planet with satellites	5.2–5.5	11.21	317.94	–1.6 to –2.94 [8]	11.86
Saturn	Gas giant planet with satellites and magnificent rings	9.5–10.0	9.45	95.18	+1.47 to –0.24 [9]	29.46
Uranus	Gas giant planet with satellites	19.2–20.1	4.01	14.53	5.9 to 5.3 [10]	84.01
Neptune	Gas giant planet	30.0–30.3	3.89	17.14	8.0 to 7.8 [11]	164.79
Kuiper Belt [12,13]	Small icy objects, a few of them several hundred miles across	30–500	0–0.18	0–0.002	+∞ to +13.6	200+
Oort Cloud	Hypothetical region, like the Kuiper Belt, but extending much further	All the way until the gravity from other systems takes over	?	?	?	Centuries+

[a]These values should be treated as approximate. Other authors quote slightly different values

Fig. 1.1 (**a–e**) The principal discoverers of the Solar System: Copernicus, Tycho, Kepler, Galileo and Newton. These men lived over a period from 1473 to 1727, more than a quarter of a millennium. The discovery of the Solar System was by no means an instant process. (**a**), Nicolaus Copernicus portrait from Town Hall in Thorn/Toruń – 1580 (Image courtesy of http://en.wikipedia.org/wiki/Copernicus#mediaviewer/File:Nikolaus_Kopernikus.jpg). (**b**) Tycho Brahe (Image courtesy of http://cache.eb.com/eb/image?id=83677&rendTypeId=4). (**c**) Kepler, 1610, artist unknown (Image courtesy of http://en.wikipedia.org/wiki/Kepler). (**d**) Galileo, portrait by Justus Sustermans (1597–1681) (Image courtesy of the National Maritime Museum, Greenwich). (**e**), Newton, portrait by Godfrey Kneller (1646–1723) (Image courtesy of http://commons.wikimedia.org/wiki/Isaac_Newton#mediaviewer/File:GodfreyKneller-IsaacNewton-1689.jpg)

Galileo and Newton. Looking back to the 1960s, when many of us first read astronomy books as a child, it is quite remarkable what scientists did not know about planets (Fig. 1.1) Some values of moon masses from the 1970s now look laughable [15]. This reflects the quality of the data, not the source.

The space probes did more to change popular perceptions of the planets than the scientists had managed in the previous 100 years. For example, the scientists very quickly discovered that life and canals could not exist on Mars, but the images of an arid desert from NASA were what really knocked this popular myth on the head. The first close-up images of Jupiter's Great Red Spot did the same.

For the first time, people could see that it was a giant whirlwind, and they did not need doctorates in science to understand what they saw.

Some nonsensical legends have proved to be harder to kill. For example: "If you could put it in a bath of water, Saturn would float." No, it would not.

First of all, the temperature of Saturn's upper atmosphere is rather colder than that of liquid water. Further, the gravitational pull of Saturn on any imaginary vat of water in some imaginary gravitational field would be significant; and finally, whatever created this gravitational field would be have to be large enough to rip Saturn apart. It would also have to be a rocky body like Earth, at such a temperature that water would be liquid. Unfortunately, Earth is the largest known body like that. The four planets bigger than Earth are gas giants, not rocky planets. The only bigger objects we know of are stars, which are gaseous and unfortunately rather too hot to sustain liquid water oceans. In short, there aren't any bathtubs to float Saturn in. (The properties of extrasolar planets are assumed from theoretical models. By great ingenuity, we have discovered quite a bit about them, and they tend to be low density objects, not high density rocky planets [16]. High density rocky exoplanets are only just being discovered at the time of writing. Who knows what will be found in the near future?)

Before printing was invented, knowledge did not diffuse through society to anything like the same extent as afterwards. The technology for copying books was a monastery full of scribes. The ancient Greeks did not even have monks. Therefore, when one of their number, Aristarchus, did discover that the planets and Earth orbit the Sun, almost no one found out. His knowledge completely failed to become mainstream.

Instead, the disastrously wrong model of Ptolemy became accepted, and even enforced on pain of severe penalties. The Church held Galileo under house arrest for years for questioning Ptolemy [17]. Ptolemy thought everything went around Earth. Once telescopes became available to astronomers, Ptolemy's theory of the universe rapidly lost what little credibility it still had.

It is often supposed that nothing further happened between the fall of ancient Greece and the time of Copernicus. The evidence does not bear out this view [18]. In fact, the data used by Ptolemy were gradually refined by medieval scholars. Copernicus knew that, and worked with better data than Ptolemy. The next great theorist, Kepler, worked with yet better data, collected by the Dane, Tycho Brahe. Galileo of course was the first great astronomer to use telescopes. Newton had data from much better telescopes, and invented the reflector telescope.

The data improved as time went on. This is an absolutely key point about astronomy. The theorists can always move faster than the observers. Theory usually catches up very quickly with new observations. The rate determining stage in making progress is almost always our ability to observe.

So what happened? Copernicus was really a refiner of Ptolemy, who had everything including Earth moving along circles-within-circles about an empty point in space [19]. Given that the world was then about as friendly to radical scientists as cats are to mice, Copernicus had the good sense not to publish until he was dying. His publisher put some conciliatory words at the front of the book to appease the ecclesiastical authorities, something to the effect that Copernicus wasn't actually telling the truth, just messing about. The ecclesiastical cat did not pounce on this particular mouse. He was soon to be dead anyway. Mind you, they didn't make this ecclesiastical canon a saint either.

Galileo declined to play the game of denying what he saw as the plain truth. Diplomat he was not. He got into big trouble, but this proved to be a Pyrrhic victory for the Church. It became plain that they had made fools of themselves, and scientists were not again persecuted until the totalitarian regimes of the twentieth century, but that is another story.

The real discoverer of the Solar System was not Copernicus but Kepler [19]. Johannes Kepler lived his life along the European fault line of the Reformation, between one round of religious wars and another. He escaped persecution, but his mother would have been burned as a witch had he not exploited his prestige as a professor to scare the witch hunters off. In other words, he lived in dangerous times.

Why then did he escape the fate of his contemporary Galileo? One of the main reasons may have been that Galileo was a very clear communicator, whereas Kepler was darn near incomprehensible. It took a man of the caliber of Isaac Newton to disentangle Kepler's writings and sort the wheat from the chaff [19]. Nobody knew that Kepler was a guy they should have felt threatened by. Even Galileo did not bother to answer Kepler's letters.

Anyway, what Kepler found out was that the planets, including Earth, go around the Sun, not in circles or circles-within-circles, but in ellipses – ovals. He also found out that the Sun is not at the center of these ovals, but at an abstruse mathematical point called the focus. An ellipse has two focuses or foci (depending whether you have read the bluffer's guide to Latin). They are not at the center of the ellipse. According to Kepler, one of the focuses of the planet's orbit is located at the Sun's center and the other is not. There is a good popular introduction to ellipses at http://www.coolmath.com/algebra/25-conic-sections/02_ellipses-intro.htm, although the planetary orbits are much more nearly circular than this diagram implies (Fig. 1.2).

Newton modified this. He realized that the Sun and the planet both orbit about a point called the center of mass (or in loose parlance, center of gravity) of the two bodies. For all planets except Jupiter, this center of mass is well inside the photosphere, the visible part of the Sun.

Kepler felt sure that there must be a simple(ish) physical reason why the planets do this. Newton found this reason: they are pulled by the same gravity that pulled the apple off the tree, allegedly onto his head.

Newton is not considered to be one of the world's three best-ever mathematicians for nothing. He showed that his laws of motion, plus his law of gravity, were enough to predict the movements of the Solar System bodies to within an accuracy that the best observers could measure. He also showed that the same is true for comets.

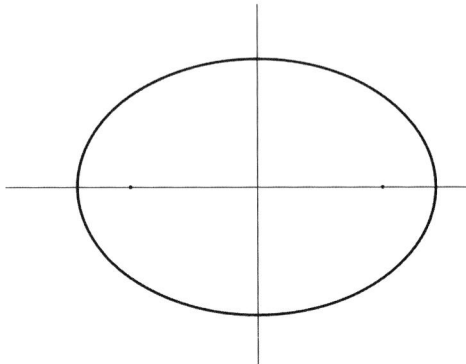

Fig. 1.2 An ellipse showing the two foci (the *dots*). When the two foci are both in the middle, the ellipse becomes a circle. In practice, the only planets whose orbits are obviously non-circular to the naked eye are Mercury and Mars. Even they have orbits much less oval than the ellipse shown (Diagram by the author)

The Naked Eye: A Stripped-Down Account

Some components of the Solar System are hard to miss. The Sun may be the only Solar System object that can be detected by blind people. You can feel its heat. Those of us fortunate enough to be sighted can easily see it. Henceforth we will assume that you have reasonably good eyesight.

The Moon is very easy to see with the naked eye.

'Naked eye' is really shorthand for 'without binoculars or a telescope.' In other words, it presupposes that you have access to glasses or contact lenses to correct your vision for focusing and eye misalignment problems. Of course, different people can see different things with the 'naked eye.' It is no consolation to the rest of us that a very few people can see the phases of Venus without optical help [20], or that our vision deteriorates with age. Vision quality is also dependent on gender. Women tend to have more color receptors than men. Henceforth, it will be assumed that your 'naked eye' is corrected with glasses or contact lenses if necessary.

If you wear varifocals, you will quickly discover that the distance-vision part of the lens has a 'sweet spot.' This is the one point at which a star looks sharp. Stars anywhere else in your field of vision are out of focus. To find this point, look straight ahead at a star and turn your head until the star becomes sharp. If you can't find this point you may need new glasses. The same phenomenon occurs with other types of glasses, but it is less pronounced than with varifocals.

Lest anyone should feel inferior because they need glasses, it should not be forgotten that even the Hubble Telescope, one of the most successful optical instruments known, needed a spectacle lens to correct a manufacturing boo-boo. When NASA finishes with it, I'm sure there would be no shortage of astronomy clubs willing to take it over. Let's face it – with that corrective lens, it's a *fabulous* telescope. So don't be ashamed of your spectacles.

Having said all this, five planets are naked-eye visible: Mercury, Venus, Mars, Jupiter and Saturn. Once a decade or so, a comet becomes naked-eye-visible for a while. Uranus is claimed to be marginally naked-eye visible, but the same question could be asked about that as about the Andromeda Galaxy. If it really is naked-eye-visible, how come the ancients did not discover it? They never had our problems with light pollution.

Nothing else in the Solar System is visible to the naked eye.

Flashlights

Astronomer's workstations are the red light district of their backyards. Why? Because your eyes' dark adaptation is not ruined by red light. It takes a good 20–30 min in the dark for your eyes to build up sensitivity. The mechanism involved is quite complex [21]. Eventually, you come to see really well in the dark, although you don't have much color vision at night.

White or blue light destroy your dark sensitivity in an instant. This is to protect your vision system from overload. You then need another 20–30 min to dark-adjust.

Red light does not affect all the dark adaptation mechanisms, so you regain your dark vision quite quickly after using a red flashlight. Even so, it is smart to point the light away from yourself, because even red lights emit some blue light. You could use bicycle rear lights. These days LEDS have replaced incandescent bulbs. They produce less unwanted blue light and use less power, so the batteries last much longer.

You can also buy red LEDs to clip onto the peak of a baseball cap. Although they free up your hands, peaked caps can be a pain at night. Yes, they can shield annoying streetlights, but they also make it hard to see the sky, and you get a lot of rude surprises when you bang the peak into a carefully aligned telescope and knock the wretched thing back out of line. That kind of thing is very easily done in the dark. The modern fashion for curved peaks also makes life difficult for varifocal eyeglass wearers. There are lens regions in varifocals that don't offer clear vision. Curved peaked caps mean that unless you adjust the cap carefully, you can only see out of the useless bits of your varifocals. Very irritating.

No doubt the fashion for curved peaks was dreamed up by some 12-year-old whiz kid who does not even know what varifocals are. But fashion is fashion. You have to have street cred.

Telescopes

Unfortunately, it's perfectly legal to be really stupid and look at the Sun through a telescope or binoculars. You will do that exactly once per eye, because you will be blinded and unable to do it again. Better still, don't do it at all. Ever. The visual receptors in your eye have no pain nerves, so you won't even feel that you have fried them. It takes no time at all.

Do you know who discovered this? Galileo, the first man to use a telescope for astronomy. He ruined his eyesight. Just to show that this danger is real, it is perfectly possible to melt a telescope eyepiece or a finder scope cross-hairs by pointing a scope at the Sun.

Portable binoculars are of limited use, except for the Moon. (In this book, Moon with a capital "M" will be used to refer to Earth's Moon, and moon with a small "m" will refer to planetary satellites in general.)

The instrument of choice is a telescope on a mount. A 6-in. (150-mm) telescope will find all the planets, and the brightest asteroids. In theory an 8-in. (200- mm) telescope will enable you to see Pluto from a dark site, but only just. Pluto and the asteroids are just dots in amateur telescopes, so you need to wait for them to move before you can identify them. They are really photographic objects.

Please stick to celestial objects, though. The story is told of a college professor who got into an interesting debate with the campus cops because the Moon was just above a girls' dorm from where he and his telescope were.

Telescopes are important enough to warrant a whole section to themselves, later in the book. We won't go through optical theory. We have no great preference among the various types. A telescope is a telescope, and all but the cheapest and nastiest will do the job.

Do resist the temptation to buy your kids inexpensive telescopes for astronomy. They really are awful, and a bad experience of stargazing can be worse than no experience. A reasonable minimum is the 3-in. (76-mm) telescopes that have recently become available. They retail for about £35–£50 in the UK and about $65 in the USA.

If you can find them, an 8-in. telescope will certainly show Neptune as a disk and a 6-in. telescope will show Uranus as a disk. They are both bluish in color. The astronomy magazines claim that they are binocular objects. If you know where to look, you can certainly see them in binoculars. But binoculars are close to useless for a first look to try to find them. You can't easily tell either planet from stars in 10×50 binoculars.

Which Constellations Are Likely to Contain Solar System Objects?

The diagrams in books usually show the Solar System as fairly planar. The planets and stuff all go round the Sun the same way, and all lie roughly on a plane.

This plane is called the plane of the ecliptic. From Earth it looks like a line going all the way around the sky. Figure 1.3 shows the constellations through which the ecliptic passes.

Not all 13 of these constellations will be visible at once. If the ecliptic is close to the horizon, as it is in your summer if you are away from the tropics, only a few of these constellations will be visible. Rather more will be visible in your winter, when the ecliptic runs high in the sky, again if you are away from the tropics.

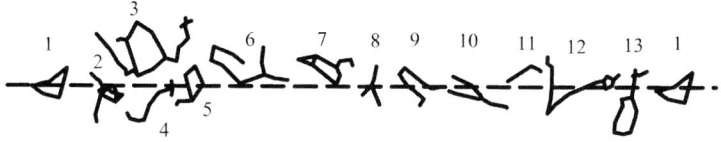

Fig. 1.3 Constellations through which the *ecliptic line* passes. Key: 1. Capricornus, 2. Sagittarius, 3. Ophiucus, 4. Scorpius, 5. Libra, 6. Virgo, 7. Leo, 8. Cancer, 9. Gemini, 10. Taurus, 11. Aries, 12. Pisces, 13. Aquarius. Capricornus is repeated to show how the *ecliptic line* runs right around the sky (Diagram by the author)

The constellations' orientations depend on where you are and the time of night. They rotate slowly through the night. They are drawn as they appear from the northern hemisphere. If you are in the southern hemisphere, these stick diagrams will appear to be upside-down to you. Of course, they do not suddenly invert at the equator. If you observe from close to the equator, they will appear to be turned on their sides compared to this figure.

British, northern Canadian and Alaskan readers are far enough north that they will not see the whole of Sagittarius and Scorpius. (Of course in summer in Alaska, the Yukon and Nunavut, it doesn't get dark enough either.) You have to be an awfully long way south in the southern summer for the corresponding phenomenon to occur. The Falkland Islands (pop. 3,100) are the southernmost populated English-speaking location at about 52°S. Castor, alpha Gemini, just about peeps over the horizon there in January. Even New Zealand is roughly antipodal to Spain, in the far south of Europe. The populated parts of the Southern Hemisphere are closer to the equator than those in the Northern Hemisphere.

The constellations through which the ecliptic passes are: Sagittarius, Capricornus, Aquarius, Pisces, Aries, Taurus, Gemini, Cancer, Leo, Virgo, Libra, Scorpius and Ophiuchus. These are where you will find planets. The asteroids and dwarf planets may stray a constellation or two north or south of these, because their orbits are inclined at a few degrees to the plane of the ecliptic. Comets can be quite out of the plane of the ecliptic, and show up elsewhere in the sky.

Broadly speaking, if you are in the English-speaking parts of the Northern Hemisphere, the Solar System will tend to be to your south. The English-speaking parts of the Southern Hemisphere extend within the tropics. If that is where you are, the ecliptic constellations may be overhead or even to your south. Outside the tropics they will be to your north in the Southern Hemisphere.

If the only constellation you know is the Big Dipper (also called the Plough or Ursa Major) or Orion, don't bother looking there for planets. The only ones you will find will be outside the Solar System; and you will need some very fancy equipment to find them. Unfortunately, you will need to learn some more constellations. At the very least, you will need to learn to recognize them with a star chart or computer screen in front of you.

Two clues with the naked-eye planets (Mercury, Venus, Mars, Jupiter and Saturn) are that they are brighter than most stars and they do not twinkle except when extremely close to the horizon. Venus and Jupiter are brighter than any stars. The planets are also visible at dawn after the stars have gone, and become visible at dusk before the stars. So if you see a "star" that doesn't twinkle, and it is in the general direction of the equator, there's a good chance it is a planet. To the naked eye, Mars is noticeably orange. Venus and Jupiter are bright white. Saturn is slightly yellow, and Mercury can look yellow or white.

The name "planet" comes from a Greek phrase *aster planetes*, meaning 'wandering star' [22]. The planets move relative to the stars. This won't help you on a single night's observing, but over a few days the movement is quite noticeable. The nearer to the Sun the planet, the more it moves. Neptune only moves by about 3° over a year, so you will have to work hard to detect its movement. With a digital SLR and with a fair bit of care and patience, you can certainly record its movement, and in principle, could even work out its orbit [23]. Uranus doesn't wander much either, but the naked-eye planets all do.

Mercury and Venus are closer to the Sun than us, so they are never far from the Sun in the sky. You only see Mercury at dawn and dusk, never in a dark sky. At high latitudes, such as in the British Isles, Mercury can be hard to find at all. Venus can be around for an hour or two after sunset or before sunrise, depending where in its orbit it is. You will not see Venus at midnight.

Mercury and Venus are called inferior planets, because they are closer to the Sun than us. All the other planets except Earth are called superior planets, because they are further from the Sun than us. This not very complex jargon does not carry any moral implications. Jupiter is not better or worse than Venus.

Maps and Software

Printed maps of the sky are fine when you are indoors, but they are not so good for outdoor use at night. They get covered in dew, you need a flashlight to see them, and you really need to hold them upside down to compare them with the sky. Holding a red flashlight at the same time, while trying not to step on whatever your kids left on the lawn that you did not see because it's dark, is not conducive to a good observing experience.

So-called planetarium software is a lot easier to use. An example is a free package called *Cartes du Ciel*, downloadable for free from http://www.stargazing.net/astropc. There are other packages that come with some telescopes, and even some you can pay for (Fig. 1.4).

It's obviously nicer if you have a laptop computer for this software, but a desktop one will do fine. Either way, take sensible precautions to keep water out of the electrics outdoors. Dewfall is not a huge problem for the computer and monitor, because they give off enough heat to stay dry. As laptops become more energy efficient, that may cease to be the case one day. For now the main electrical issues are the connections and the wiring.

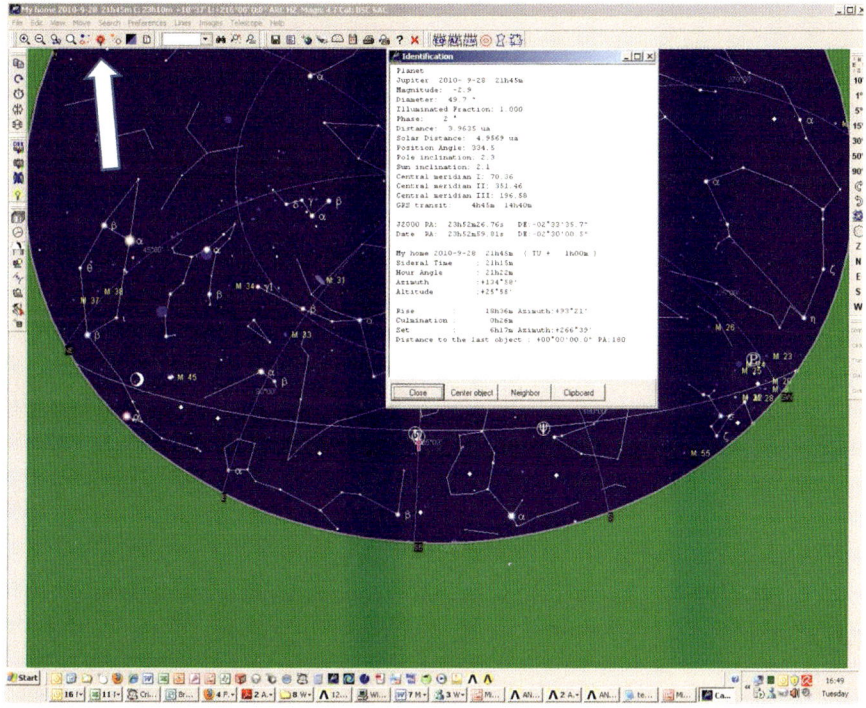

Fig. 1.4 A screen dump from Cartes du Ciel. The inset window contains the information you get about Jupiter when you click on it. Notice the *red* light bulb icon (*arrow*) (Image by the author)

Cartes du Ciel runs on Microsoft Windows™. If you prefer a different operating system, you will need to make sure it has a way to run Windows software. The *Cartes du Ciel* website says it runs on versions of Windows up to XP. It can be run under Windows Vista without problems.

One of its nicest features is the red light bulb icon. If you click on this, your whole computer switches to red-based colors. This makes it feasible to use this software outside in the dark. Your computer screen is more or less vertical, and the stars look the same way up as you see them in the sky. You can use dead time while exposing photographs etc. to sit and learn the sky from *Cartes du Ciel*. It is well worth the effort. Do also check how to dim your computer screen. Practice this in the light. Learning stuff in the dark on a cold night is always that bit harder (Fig. 1.5).

However, while you build up your knowledge, *Cartes du Ciel* will plot the ecliptic line if you ask it to, and will show the planets. It knows the time and date and will follow the movements of the planets. You can download add-ons to show asteroids. There are squillions of these, but it is inadvisable to display more than 20; otherwise you'd never see anything.

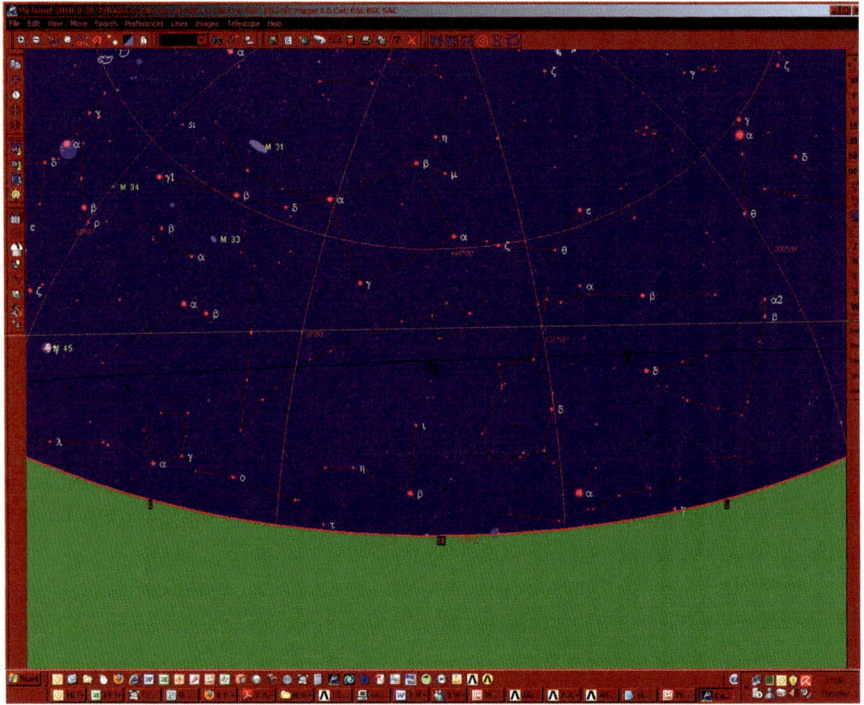

Fig. 1.5 A screen dump from Cartes du Ciel after the *red light bulb* has been clicked. Notice how Windows switches to *red*-based colors. It's a pity there is still a *green* horizon (Image by the author)

You need to program in your location using longitude and latitude. You can get these from www.muiltimap.com. Type in your address, and it will give you your longitude and latitude. This website works for the United Kingdom, the United States, Canada, Australia, New Zealand and South Africa. Once you have done this, *Cartes du Ciel* defaults to the time and date stored in your PC. You can override these to find out where Mars will be next week or whatever (Fig. 1.6).

You can even put in the dates when Galileo first observed the satellites of Jupiter, and reproduce the diagrams in his book. As it happens, he timed his observations to take place before the Moon rose. Presumably this was not accidental.

This is not the place for a *Cartes du Ciel* tutorial. The software is rich with capabilities and easy to learn (Table 1.2).

The naked-eye planets do not twinkle like stars. Their glow is constant as long as they are well clear of the horizon, where not even the Sun is undistorted sometimes.

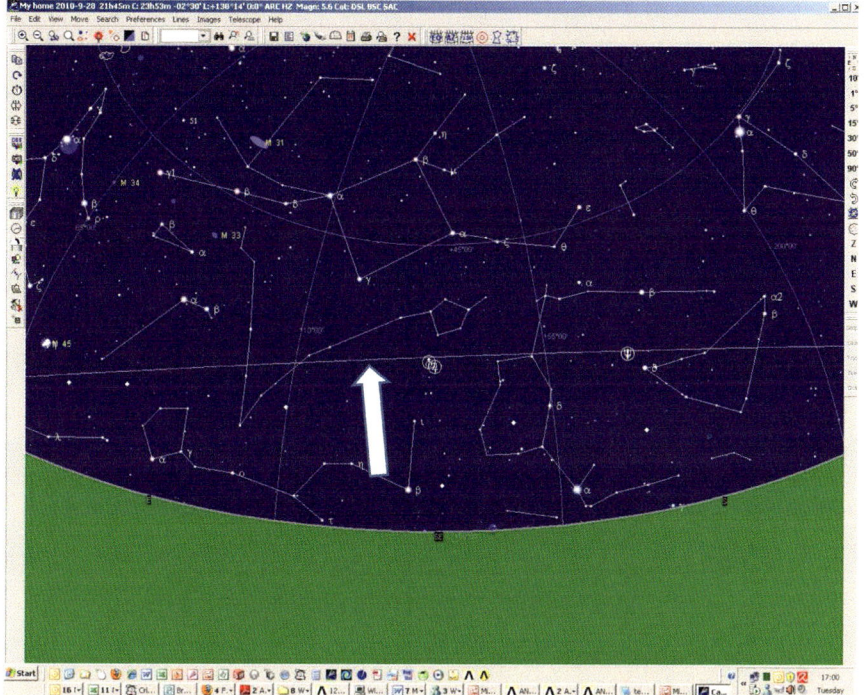

Fig. 1.6 A screen dump from Cartes du Ciel showing three planets along the *ecliptic line* (*arrow*). Three planets are shown by their symbols, which are listed in Table 1.2 (Image by the author)

Table 1.2 The symbols used for the planets. These mostly have ancient origins, although since Pluto was not discovered until 1930, its symbol is obviously not ancient. Since it is common, the comet symbol has been added

Planet	Symbol
Mercury	☿
Venus	♀
Earth	⊕
Mars	♂
Jupiter	♃
Saturn	♄
Uranus	♅
Neptune	♆
Dwarf planet Pluto	♇
Dwarf planet Ceres	⚳
Comet	☄

What Not to Wear

Fashion is not normally emphasized at star parties and other gatherings of astronomers, unless you call not caring how you look a fashion statement. Avoid anything that needs dry cleaning, or you may be in for an expensive surprise. If you polar align your mount, you spend enough time kneeling at your telescope to justify investing in a gardener's foam mat for kneeling on. It hurts less and your knees stay dry. You may even sit on this mat to adjust your mount, because sitting on the ground is often the only way to see through the little polar-aligning telescope. You will regret wet knees, seat and feet if the weather is cold.

If you are serious about your dark adjustment, especially if you have to observe near street lights, you may wish to consider darker clothing. A bright colored jacket worn for cycling would completely prevent you from getting dark adjusted. You will not need to try that twice. You can buy waterproof trousers[1] from bicycle shops. They are not warm enough to wear on their own, but as an over garment there is a lot to be said for them if you are outside for any period of time observing on either wet ground or snow. They also keep the wind out. Trust me, after three or four hours outside on a cold, crisp night, you will get really cold.

Clear nights are invariably cold nights, because there is no cloud to reflect heat back to Earth's surface. Wrap up warmly because you will not be moving about much. Wear a good coat, hat and scarf. Layers of clothes tend to work better than single, thick garments in my experience. You would do well to have a pair of fingerless gloves with removable flaps to cover the fingers and thumbs. On the coldest nights you could put a pair of large mittens over the fingerless gloves during periods of little activity.

If you are outside for long periods exposing photographs, you will probably need to move about to stay warm, but don't do your exercises where you will make the telescope vibrate. Vibrations get magnified just as much as your image. Even if you wear two pairs of everything: trousers, socks, two sweaters – you can and still will get cold.

You cannot wear the coats you buy in the North American Frost Belt indoors. They are just too warm. You cannot say that about most coats bought in Britain. If you live in a region that does not get snowy winters, and visit such a region, you could think about buying a coat for astronomy. One thing a purpose-bought coat needs is pockets – lots of them, preferably without fiddly zip fasteners you cannot operate with one gloved hand. It's harder to lose an eyepiece in the dark if it's in your pocket than if it is on a table or on the grass.

---

[1] In British English, 'pants' means an undergarment. It is never used for outer garments, which are always called trousers. 'Trousers' in American English carries overtones of very fancy garments that only a diplomat would wear. In British English this is not so. You cannot avoid sounding slightly comical on one or other side of the Atlantic. 'Trousers' have been opted for on the grounds that this at least avoids lavatory humor.

You should wear boots with thick socks. So-called moon boots, i.e., non-waterproof boots with foam lining, are undoubtedly warm, until the dewfall on the grass makes them wet. Then they become useless. Rubber boots, affectionately known in Britain as wellies, after their inventor the Duke of Wellington, work quite well in mild winters, provided you wear a couple of pairs of socks. Fur-lined wellies are no better than wellies-and-socks. In those parts of North America that get cold winters, good snow boots are *de rigeur*. Hiking boots are not uncommon among astronomers, but have the disadvantage that you can't quickly slip them on and off as you enter and leave your house. High heels are obviously not the best idea in the dark, especially as they sink into soil. Besides, in the dark, you trip over everything, and don't need twisted ankles (Fig. 1.7).

In some parts of North America it gets too cold for manmade fibers. If that applies to you, then you will presumably possess suitable clothes.

If you are in sunny California and make a trip to the mountains to get above the smog, don't forget that it can get seriously cold up there at night, and dress accordingly. Mountains can also be much windier than low-lying areas. Consider the wind when you choose where to site your telescope. If you plan to go really high, it makes sense to take the usual mountaineering precautions about having someone know where you are and when you are expected back, take a tent and food in case the weather suddenly turns nasty, etc. You should also of course wear proper mountaineering clothes. Also, if you go up a mountain, and your car is not normally started in the cold, do have a mechanic check beforehand that your car will start in the cold at low air pressure. The mountains are no place to have a breakdown! The reward for the effort is that, high up, the skies are just astonishingly clear.

On summer nights, you get another hazard – insects. The little horrors will eat you alive given half a chance. You need to cover up well. You can of course invest in insect repellent sprays and creams. They work, but aphrodisiacs they are not. Many brands stink and are a pain to wash off. Insects often prefer the evening and dusk to dead of night. They are also much more numerous near trees and hedges.

Do you enjoy being afraid? You needn't watch horror movies. Just sit outside in your backyard in the dark and wait for the animals to come to life. Cats are the least of your worries. Apart from the odd yowl when they get randy or scream when they attack one another, they are professionals at keeping a low profile. Most small animals are also discreet because they are avoiding the cats. There is one animal that is not the tiniest bit scared of cats: the cute, prickly hedgehog. No mammal or bird is going to attack it because it won't come off best. Consequently hedgehogs do not care a whit how much noise they make. For creatures so small they sure know how to make a racket, sniffing among dead, crackly leaves. Most astronomers have their share of scare stories, but also stories of animals that are sociable enough to keep them company at night.

Owls, too, are normally the soul of discretion, unless you frighten them. An owl making an emergency exit gets its revenge on you for frightening it. You can't see it, but all of a sudden you are terrified by this tremendous kerfuffle. Score: one all. Seeing one fly overhead at night is quite a sight. They are big, graceful and utterly silent.

Fig. 1.7 My baby and me. I use this 6-in. Newtonian to guide my DSLR, which is piggybacked on via the semi-homemade mount shown. Mostly, however, I wanted to comment on my dubious attire. The boots are waterproof. At night I wear them with socks. Everything else is machine washable – it can get very muddy. The coat has plenty of pockets. The gloves are OK for autumn, when the picture was taken. In winter I use fingerless ones with a flap to cover the fingers. The clothes are also rugged. You bump into and scrape all sorts of twigs and stuff in the dark, which would ruin more delicate clothes. Any earrings I might lose in the dark have been removed. My hair is kept out of the way with a hair band. If it were really cold I would wear a hat. I do not favor baseball caps. You bump into too many things with their peaks as you try to look down eyepieces, finder scopes, etc. In summer I would cover my legs, neck and head against insects (with permission from Maggie Hathaway-Mills)

Where this Book Goes from Here

Subsequent chapters will show you how to get the best from observing the various Solar System objects. You will learn what equipment you might wish to think about. You don't have to buy everything suggested. Many astronomy clubs have telescopes and other equipment for loan. You can then try equipment out at minimal cost, and work out whether the investment is right for you.

One of the advantages of Solar System astronomy is that you can see and photograph a lot with relatively inexpensive equipment. It is by no means essential to spend 'big bucks.'

Chapter 2

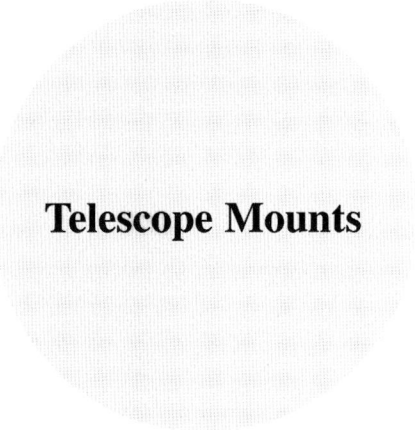

Telescope Mounts

What a Mount Does

What a Mount Does

Figure 2.1 shows how a celestial object that we may wish to observe does not stay in the same apparent place in the sky. It seems to drift westward along an arc, and it appears to rotate. This apparent rotation is known as *field rotation*. We all probably know that both the westward drift and the field rotation are illusions: it is really Earth that is rotating. It rotates so slowly that we do not sense the motion.

The purpose of a telescope mount is firstly to save you the impossible job of holding a sizable telescope still, and secondly to enable you to keep pointing the telescope at an object which moves relative to the ground.[1] All comments made in what follows also apply to a mount for binoculars.

The mount has to take a telescope, and move it precisely while it is being buffeted by wind of unknown speed, which tries to make the telescope flutter like a flag or twist like a weathervane. Any imprecision is magnified by the telescope along with the image. If you are using a 100× eyepiece or camera, the vibrations and image drift are multiplied by 100.

You therefore need a mechanism that combines stiffness with precision. This does not make for an easy engineering design problem. It's like expecting a champion sumo wrestler also to be a top ballet dancer.

---

[1] The information in this chapter comes from many sources. Some of it was learned by looking at peoples' instruments and talking to use, some from books, some from the internet, especially from reading manuals for mounts, and some from plying my trade as an engineering physicist.

© Springer Science+Business Media New York 2015

J. Clark, *Viewing and Imaging the Solar System*, The Patrick Moore
Practical Astronomy Series, DOI 10.1007/978-1-4614-5179-2_2

Fig. 2.1 Celestial objects appear to move across the sky during the night. As they do so, they also appear to rotate. Of course this is an illusion. It is really Earth that is rotating, but we do not sense this (Image by the author)

In popular parlance, the terms stiffness and strength are freely interchanged. Engineers and scientists use stiffness to mean ability to resist deformation, and strength to mean the load or force required to break something. Do you remember that biblical phrase "out of strength came forth sweetness?" It relates to the time Samson took honey from a dead lion. Somehow, though, "out of the stiff came forth sweetness" would not have the same ring, even if the lion actually was a stiff at that point. Not so for astronomers. For us, out of stiffness, sweetness really does come forth.

How well you need to solve this stiffness-and-precision conundrum depends on what you want to do. If you just want to look through the eyepiece, you can get away with a much weaker, and therefore less expensive, mount. This is because your eye and brain are remarkably efficient at correcting for a shaky image. Also, you probably don't that much mind losing a few seconds observing if a strong gust of wind passes by.

If you are trying to draw what you see, you probably want something a little stiffer.

Cameras, however, record every little shake and tremor. They can't filter them out like a brain can. Every one of them appears in the finished image. Therefore you need a mount stiff enough that it does not deflect, bend or move for the duration of a single exposure. Not quite: the mount has to move to counter the earth's rotation. You don't want any unwanted movements: movements that do not contribute to pointing the telescope at the same area of the sky.

This is a more difficult problem than making a telescope. The optics in your basic Newtonian or refractors were sorted out a long time ago. Most of their advantages are that they compensate for the inadequacies of mounts, at least at the amateur level. Professional telescopes are, of course, a different matter, but amateurs have much smaller budgets.

How Do Mounts Rotate to Follow the Sky?

The first concept we need is that of a rotating joint. Mount structures are made of stiff metals like steel or aluminum, which are anything but bendy. To allow rotation, we therefore need joints. The simplest rotating joint is the so-called revolute joint.

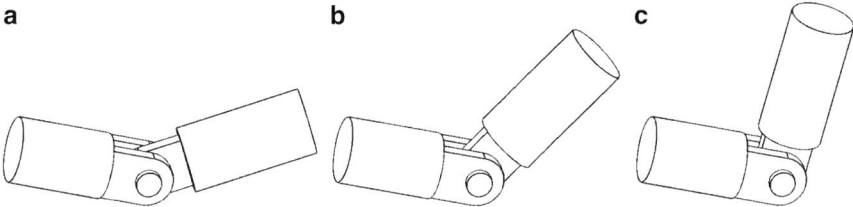

Fig. 2.2 A revolute joint will rotate in one direction only. Examples of revolute joints are door hinges and (approximately) elbows and knees (Image by the author)

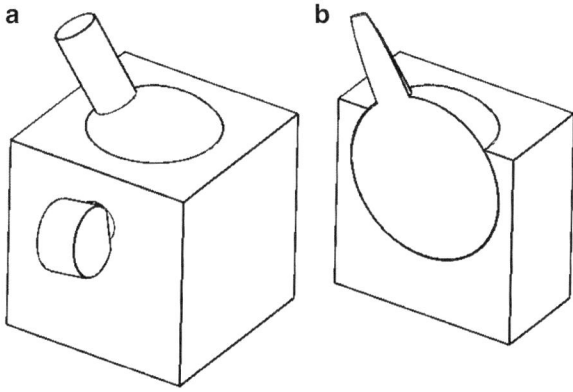

Fig. 2.3 A ball-and-socket joint can swivel in any direction. The left-hand figure (**a**) shows how a thumbscrew is often used to fix the joint once it is rotated to the desired position. The right-hand figure (**b**) shows a section view showing how the ball and socket fit together (Image by the author)

A telescope mount that only rotated in one direction (Fig. 2.2) would be of pretty limited use. That is not to say that such telescopes have not been used. Lord Rosse used a telescope with declination adjustment only, and he was the first to discover spiral structure in a galaxy [24]. Not too shabby.

It would be much more practical to have a mount that can swivel in more than one direction. Ball-and-socket camera mounts (Fig. 2.3) were somewhat popular 40 years ago, but they are less common today.

Although it was extremely easy to orient the camera to any chosen direction, it was very difficult to get the thumbscrew to tighten sufficiently to make the joint stiff. From an astronomical point of view, it is also not obvious how to drive such a joint to follow the sky as Earth rotates. We have ball-and-socket joints in our hips. To move our legs, they rely on muscles attached to the bones by ligaments. We can stand upright for quite long periods, but we can't hold our thighs at other angles and hold our body weight for more than a few seconds. So they are not a good weight-bearing joint. We also use ball-and-socket joints to hold our eyes while our muscles rotate them. The eyes are almost spherical; holding their weight at any given angle is not difficult. But even they are not true ball-and-socket joints. We cannot turn our eyes upside down in our skulls. They will not rotate about the axis through the pupil. (For our eyes to be able to rotate this way would present mother nature with some interesting design challenges, such as how to prevent the blood vessels and optic nerves being damaged by twisting.)

The trouble with a telescope is that it needs the maneuverability of our eyes but has a weight distribution more like our legs. We need a different solution to the orientation problem.

Altitude-Azimuth Mounts

The solution employed by nearly every mount known is to use multiple revolute joints.

The simplest two-axis mount commonly used by amateur astronomers is the altitude-azimuth mount, commonly abbreviated to alt-az. An example of an alt-az mount is shown in Fig. 2.4.

This is by no means the only possible design for an alt-az mount, but it is the most common, and is sometimes called a 'fork mount.' The component for which it is named, the fork or clevis, is shown in Fig. 2.5.

Notice how the clevis drawn is rather tall. If it were not, the telescope shown in Fig. 2.4 would bump into it when it is pointed at high altitudes.

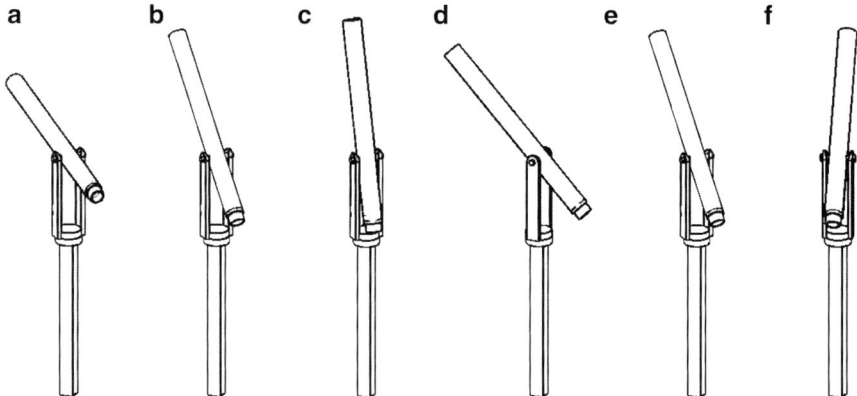

Fig. 2.4 A telescope in an altitude-azimuth or alt-az mount. The upper three pictures show movement in the altitude (*up-down*) direction. The lower three show movement in the azimuth (*left-right*) direction. In all cases, we have shown the mount attached to a single leg. This arrangement is common in permanent observatories. It is much more common for portable mounts to sit on a tripod (Image by the author)

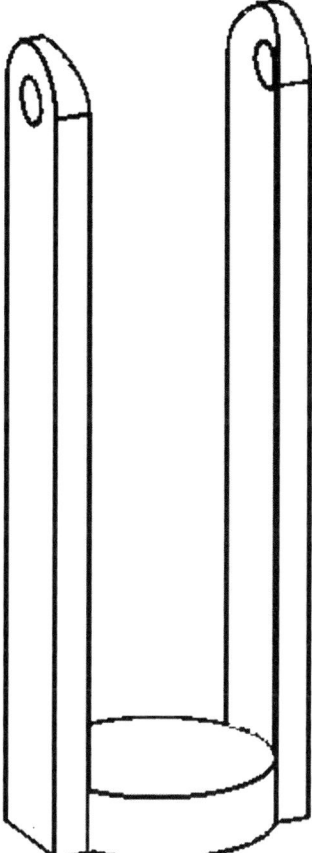

Fig. 2.5 A fork or clevis, the component for which fork mounts are named (Image by the author)

This mount actually has a blind spot when pointed at the zenith, the point in the sky that is directly overhead. A Newtonian telescope, in which the eyepiece of the telescope is at the side, not the back, would not have such a blind spot.

This fact was exploited in a design of telescope called a Dobsonian [25]. Dobson's idea was to cut the cost of a simple telescope as much as possible. He made many cost-cutting simplifications to Newtonian telescopes, but the one that caught on was the mount. A Dobsonian telescope is shown in Fig. 2.6.

This ingenious invention relies on two things. First, the center of gravity of a Newtonian telescope is quite close to the mirror, because that is by far its bulkiest item. Second, the mount is especially suited to telescopes whose lengths are approximately the same as the height of an adult human. Thus the eyepiece would normally be somewhere around head height.

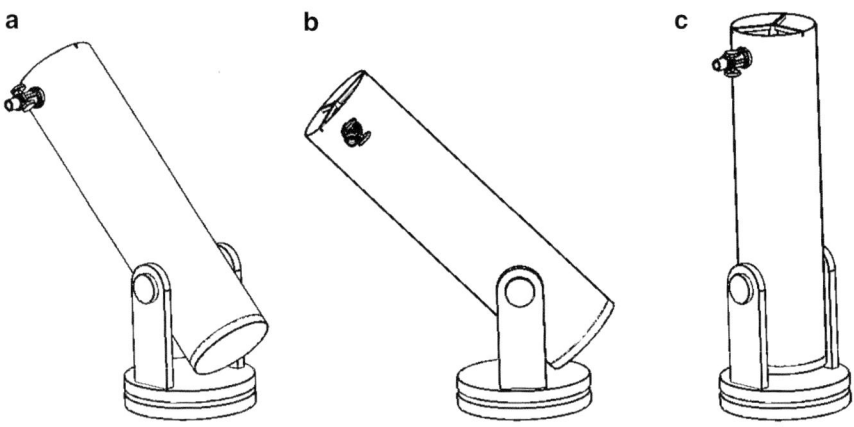

Fig. 2.6 A Dobsonian telescope consists of a Newtonian optical tube assembly (OTA) mounted on a turntable that sits on the ground. The upper part of the turntable rotates in the azimuth direction. It, in turn, holds the telescope on pins mounted to coincide with the center of gravity of the OTA, which allows the 'scope to rotate in the altitude direction. Notice that there is no blind spot at the zenith (Image by the author)

The original idea was that a Dobsonian mount would not be motorized, and that the user would move the telescope to track the sky. It was meant to be a way to make large Newtonian telescopes affordable for everyone. In this, the idea was very successful. The mainstream telescope manufacturers soon adopted the idea, and produced some very lightweight designs with open tubes made from trusses. It is also a mount that can be made by do-it-yourself enthusiasts. Some amateur astronomers have produced very fine examples.

However, it was not long before motorized versions appeared, some with GOTO capability and with apertures up to at least 23 in. You have to stand atop a tall ladder to look, which is certainly not for the faint-hearted. The really large 'Dobs' were a fad, which did not last. However, you can still buy a new 16-in. instrument.

For what kind of observing are these telescopes good? A non-motorized telescope is good for looking down, otherwise misleadingly known as visual observing,[2] but it is not really suitable for drawing or photography. For these activities you really need a motor drive.

The challenge when motorizing an alt-az mount is that it has to move in both the altitude and azimuth directions to track a celestial object, and the required amount of movement is different for every object. You need a microprocessor-driven drive system. Such systems are obviously not the cheapest, but they are not outrageously expensive nowadays.

There is another important challenge, which only really matters for long-exposure photography: these mounts do not follow field rotation. Five minutes per frame is about the exposure limit.

Let's turn our attention now to a mount that can follow field rotation.

---

[2]What other kind of observing is there besides visual? Listening to your images? Also, if photography is not a visual medium, what is it?

Equatorial Mounts

Like alt-az mounts, equatorial mounts work by combining revolute joints to give rise to two-axis motion. The difference is that instead of aligning the axes of these joints with the vertical and horizontal, one axis is aligned with Earth's axis of rotation. If you then make the telescope rotate about this axis in the opposite direction to Earth's motion, it will always point at the same part of the heavens (Fig. 2.7).

The declination axis is then only used initially to point at the observed object. The mount then does not need to move about this axis. More accurately it would not if it were perfectly aligned. In practice, except in the highly unlikely event that you align the axis perfectly, you will need to make fine corrections as your mount tracks an object.

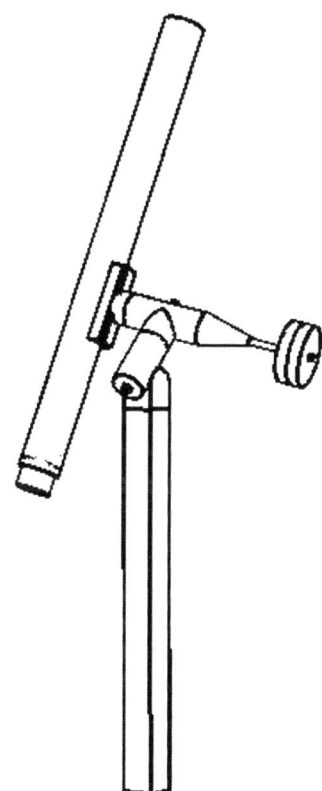

Fig. 2.7 An equatorial mount has one axis aligned with Earth's axis of rotation (Image by the author)

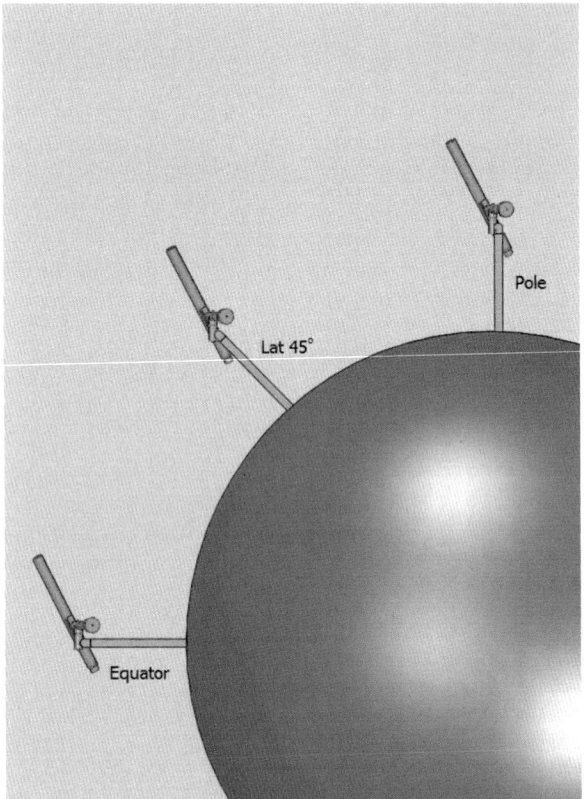

Fig. 2.8 Three identical telescopes with equatorial mounts, each pointing at the same celestial object but from different places, each having the same longitude but one at the equator, one at the pole and one halfway in between. Notice how, although each stands on a (locally) vertical pillar, they all have an axis aligned with Earth's axis of rotation (Image by the author)

You don't need a microprocessor to do this with a motor-driven mount. This kind of mount was pretty well the only motorized option for amateur astronomers before the computer age. Many can still remember wind-up clockwork drives on telescope mounts. Nowadays electric motors are much more common. Most come with hand controllers that allow you to fine-tune the pointing. The more expensive motorize both the right ascension and the declination movement. Cheaper ones may only motorize the right ascension movement, leaving a hand control for declination movement. The hand control is often a knob mounted on a rubber stalk, to make it easier to use.

These mounts contain two more revolute joints to enable the user to align the axis of one of the revolute joints with Earth's axis of rotation (Fig. 2.8).

Aligning an Equatorial Mount

Equatorial mounts actually have a built-in alt-az mount to allow this to happen. First, get your tripod or pedestal as level as possible. It is much easier to do this in daylight – spirit levels are a pain to use at night. A useful tip is to align the spirit level with each of the legs of a tripod so that you can see which way that leg needs to be adjusted. Then put the full weight of the telescope and the counterweights on so that the mount can settle for at least an hour. At a pinch you can omit this step if you are not at home, but do not omit it if you are at home. Depending how hard your ground is, which in turn depends on the weather and the season, you will have to repeat this step for a tripod on your lawn between once a week and once a month. A pedestal mounted in concrete should hopefully never need re-aligning. Most mounts can be re-leveled without removing the pedestal from the ground.

Second, roughly align the mount along a north–south line. Again, this is easier done in daylight if at all possible. If you use a compass, remember that the steel in the mount will make it point in the wrong direction. You will easily discover by experiment how far away the compass needs to be.

The fine alignment has to be done at night. Do not take too seriously the scales on the altitude axis purporting to tell you the latitude.

In the Northern Hemisphere it is easy to find north at night. There is a second-magnitude star very near the north celestial pole called Polaris. In the Southern Hemisphere it is quite a bit more difficult, especially in light-polluted areas. There are stars close to the south celestial pole, but they are dim. The brightest nearby stars are fifth-magnitude stars in the constellation of Octans. You are not going to see these with the naked eye in a street-lit town.

Third, tilt the azimuth mount until it points roughly at the celestial pole for your hemisphere.

Now you need a way to fine-tune the adjustment. For inexpensive mounts you are frankly reduced to doing the best you can. They have no polar finder scope. You could do worse than to set your latitude using the scale on the mount, and placing a set square against the tube covering the declination axis, because the thin metal part of the set square will make a good right angle with the declination axis. Then look along this like a gun sight. If you are middle aged or older this is not easy: you will not be able to focus up close. But it is possible, and does a tolerable job good enough for visual observing. Furthermore, such telescopes are quick to set up and suitable for taking to your friends' houses or astronomy club gatherings.

The next best trick, which works well in the Northern Hemisphere, is to have a polar finder scope. This is a small telescope that sits in the mount and is placed on the right ascension axis (see Figs. 2.9 and 2.10). This finder scope has a cross-hair that you align on the pole. Quite often, it will also have a little ring around this cross-hair, with a small circle on it (Fig. 2.11). The purpose of this circle and ring is described in the caption for Fig. 2.11.

Polaris is not quite at the north celestial pole. Its declination is actually about 89° 16'. It is about three-quarters of a degree away. If it were at the celestial pole, its

Fig. 2.9 A polar finder scope may be inserted along the right ascension axis of an equatorial mount (Image by the author)

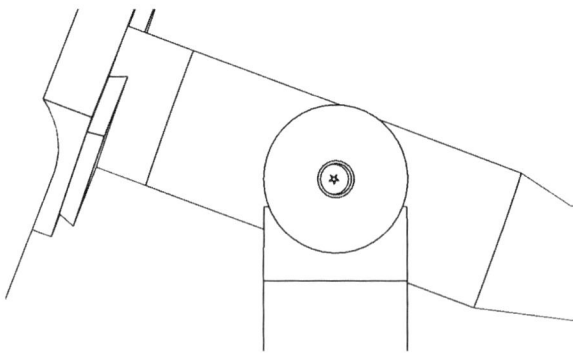

Fig. 2.10 A polar finder scope inserted along the right ascension axis should be pointed at the pole star for your hemisphere (Image by the author)

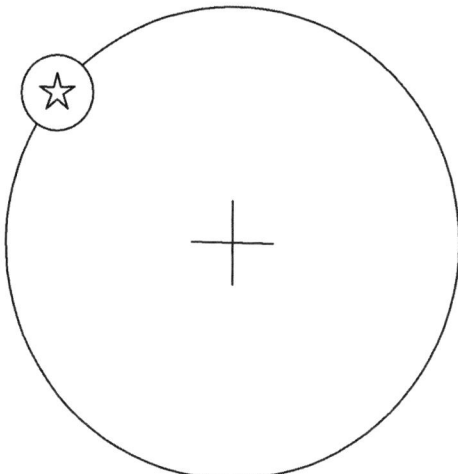

Fig. 2.11 The view through a polar finder scope. The little ring on the *circle* should be moved to the correct orientation for your time and date. The pole star for your hemisphere should then be centered in the little ring by adjusting the altitude and azimuth settings on your mount. The cross-hairs should then be fairly accurately centered on the celestial pole. This is not a perfect process. The same ring is not valid for Polaris and Polaris Australis. Also, the orientation of Earth's axis of rotation is not quite constant (Image by the author)

right ascension would be meaningless. This is for the same reason that a house at the North Pole would have all four walls facing south – the lines of longitude all meet there. Similarly all lines of right ascension meet at the celestial poles. As it is, Polaris does have a right ascension; its J2000 RA was 2 h 31 m 48.70 s. That close to the pole, the proper motion does not need to be very great for the RA to change rapidly. In addition, Earth's axis of rotation precesses (i.e., wobbles), making a complete wobble every 27,000 years. Isaac Newton showed this phenomenon to be due to the slight oblateness of Earth. It is not quite a sphere, and the Sun's gravity picks up this slight out-of-roundness and uses it to drive precession. You can look up the present position of Polaris at http://arxiv.org/PS_cache/astro-ph/pdf/0002/0002406v2.pdf [26].

The ring and circle are set for the position of Polaris at a particular time, presumably fairly close to the time of manufacture. The idea is to place Polaris in the ring shown in Fig. 2.11.

The correct orientation for the little circle depends on what time of the sidereal day it is, or if you don't know this, the regular time and the date. Many mounts have a set of rings around the eyepiece of the finder scope to enable you to translate the date and time into the correct RA for Polaris, and hence the orientation of the ring and circle. If you bought a second hand mount with no manual, you can usually download the manual off the Internet. (You can also do this for everything from your central heating controller to your washing machine.)

Some manuals also have images of the constellations Ursa Major and Cassiopeia to enable you to align your RA axis with these. This can be an awkward and tedious process, not least because no one ever seems to make polar finder scopes with long eye relief eyepieces for eyeglass wearers. Manufacturers please note: this is a hint.

For Southern Hemisphere readers, the process is that four stars of the constellation Octans – upsilon, tau, sigma and chi – are sometimes engraved into the ground glass at the focal plane of the finder scope. You align these stars in the circles placed there. They are all fifth-magnitude stars, so this is going to be a pain in a light-polluted area. Sigma Octans is sometimes known as Polaris Australis. Its declination is almost exactly 89 °S.

The state of the art process for aligning amateur scopes is drift alignment. In principle this is straightforward, provided you are not worried about the geometry behind the technique and are happy to follow a recipe. It is a little time consuming, and only worth the bother if you are attempting photography or drawing.

Anyway, the recipe is as follows. You can either use an eyepiece with cross-hairs or a digital camera that is capable of producing images live on a laptop screen.

You don't need any eggs, flour, milk, a whisk and frying pan, but a good idea of where your local meridian is, the imaginary line from your zenith going due south. It also helps to know where your celestial equator is. Look on a star map or a program such as *Cartes du Ciel*.

The drift alignment procedure is given first for the Northern Hemisphere, then for the Southern.

Northern Hemisphere Drift Alignment

Align your tripod with the north celestial pole as well as you can. Point the telescope towards the celestial equator and due South. Pick the nearest reasonably bright star.

Turn the camera or the eyepiece so that one of the cross-hairs points along the direction in which the star 'moves' if the motor drive is momentarily disconnected or switched off. This direction is west. The opposite direction is obviously east. You will get less confused if you rotate the camera so that West is left and east is right, just like on a map.

Which way is north now on your computer screen? The answer to this question is unfortunately not obviously 'up.' It is up if your telescope is Newtonian or if it is a refractor with no right-angle bend in the optical path to the camera. If your telescope is a refractor or a catadioptric with a right-angle mirror in the eyepiece, north will now be down.

Paradoxically, we now use north–south drift to align the mount in the west–east direction. If the star drifts to the north (hopefully slowly), move your azimuthal adjustment on your mount east; if the star drifts to the south, move it west. Repeat until the star no longer drifts either north or south. Do not worry if it drifts either east or west – you can sort that out later. Every time you move the mount, you may have to select new stars.

Once you have got the north–south drift reasonably under control, correct for east or west drift. First, point your telescope at the eastern horizon, and look for a star reasonably near the celestial equator. If this star drifts north, adjust your altitude (latitude) setting to the south and vice-versa, until the drift ceases.

Repeat the last two steps as often as needed. You should hold the stars still in your view for as long as you wish to expose a photograph. If this means shooting footage with a webcam for 5 min, hold the image still for the full 5 min if you wish to avoid field rotation.

Southern Hemisphere Drift Alignment

Align your tripod with the south celestial pole as well as you can. Point the telescope towards the celestial equator and due north. Pick the nearest reasonably bright star.

Turn the camera or the eyepiece so that one of the cross-hairs points along the direction in which the star 'moves' if the motor drive is momentarily disconnected or switched off. This direction is west. The opposite direction is obviously east. You will get less confused if you rotate the camera so that west is left and east is right, just like on a map.

Which way is now south on your computer screen? The answer to this question is unfortunately not obviously 'down.' It is down if your telescope is Newtonian or if it is a refractor with no right-angle bend in the optical path to the camera. If your telescope is a refractor or a catadioptric with a right-angle mirror in the eyepiece, south will now be down.

Paradoxically, we now use north–south drift to align the mount in the west–east direction. If the star drifts to the north (hopefully slowly), move your azimuthal adjustment on your mount east; if the star drifts to the south, move it west. Repeat until the star no longer drifts either north or south. Do not worry if it drifts either east or west – you can sort that out later. Every time you move the mount, you may have to select new stars.

Once you have got the east–west drift reasonably under control, correct for north or south drift. First, point your telescope at the eastern horizon, and look for a star reasonably near the celestial equator. If this star drifts north, adjust your altitude (latitude) setting to the south and if it drifts to the south, adjust your altitude (latitude) setting to the north, until the drift ceases.

Repeat the last two steps as often as needed. You should hold the stars still in your view for as long as you wish to expose a photograph. If this means shooting footage with a webcam for 5 min, hold the image still for the full 5 min if you wish to avoid field rotation.

Mounts have been quite deliberately discussed at some length before telescopes and binoculars, because the mount is a more difficult piece of equipment to get right.

In the next chapter, we will look at telescopes and binoculars.

Chapter 3

Telescopes, Binoculars and Light

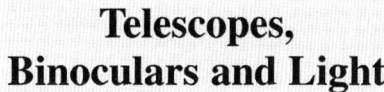

Electromagnetic Radiation: Our Window on the Universe

Just about every observation we ever make of the Solar System is made using electromagnetic radiation of one form or another. The differences between the forms are differences of degree, not differences of kind. Exactly what electromagnetic radiation is was a puzzle from at least the time of Isaac Newton until the mid-twentieth century.

The picture that emerged in the first half of the twentieth century, the quantum theory, was finally able to explain why electromagnetic radiation sometimes has particle-like and sometimes has wave-like properties. The quantum picture of all sorts of 'particles' is one in which there is a particle, but the position of the particle is not deterministic. We can only make statements about the probability of detecting the particle at any point in space and time. The probability is governed by a wave function, an entity that obeys various wave-like laws. The name for a particle of electromagnetic radiation is a photon.

One of the first properties to be discovered was that the energy possessed by a given photon is inversely proportional to its wavelength, i.e., the wavelength of its associated wave function.

Many popular expositions of quantum theory are available. Unfortunately anyone who reads them is in for disappointment. Anyone with a high school education can understand the words in Newton's laws of motion, even if they find the concepts a little puzzling. The blunt fact is that you need a few years of the kind of mathematics taught to physical scientists at a university to understand the words and equations used to write down the laws of quantum mechanics. For example,

© Springer Science+Business Media New York 2015
J. Clark, *Viewing and Imaging the Solar System*, The Patrick Moore
Practical Astronomy Series, DOI 10.1007/978-1-4614-5179-2_3

you need to know that observable quantities are assumed to be eigenvalues of Hermitian operators. Don't panic, though. All that will be skipped in this book.

(If you have the time and energy, you could look at Riley [27] for the mathematics, an out-of-print but still obtainable book by Littlefield and Thorley [28] for the experimental evidence and Gasiorowicz for the quantum theory [29]. That will get you to an elementary level. To proceed further, you will need someone like Muirhead [30] for the theory of relativity, also out of print but obtainable and much the best book of its kind, then Strange [31] for relativistic quantum mechanics. Most bachelors' degrees in physics do not take you that far.)

Our current picture of nature has four types of 'force,' electromagnetic, gravitational, weak nuclear and strong nuclear [32]. Of course physicists would love to have a grand unified theory that unites them all. There has been some success. The relation between electromagnetic and weak nuclear energy is known. The main effort at unification, so-called string theory, suffers from a key weakness – it is totally untainted by anything quite so nasty as evidence. It is likely that, once the relevant evidence emerges, string theory will require significant modification. Indeed, the recent discovery of a repulsive gravitational force causing the expansion of the universe to accelerate [33, 34], suggests that we may not have discovered all the forces of nature.

The intermediary particle for electromagnetic force is the photon [35]. For gravity it is hypothesized to be the graviton, but no one is even close to observing one of these. For the weak nuclear force the intermediaries are the W and Z bosons [36], and for the strong nuclear force the intermediary is the gluon. W and Z bosons have been observed in enormous particle accelerators. Gluons can only exist within the tiny world of atomic nuclei. (If an atom were the size of your nearest cathedral, its nucleus would be the size of a pea.)

Photons, on the other hand, are stable for billions of years and can travel across the known universe. They are therefore our primary means of observing the Solar System, and certainly the only means accessible to amateur astronomers.

We use electromagnetic radiation to infer the actions of the other forces.

Figure 3.1 shows that Earth-based astronomers can only observe at certain wavelengths, those to which the atmosphere is transparent. Visible electromagnetic radiation is no more and no less than the stuff we call light. We can observe a little way into the ultraviolet and infrared regions. An amateur's digital astronomical camera is certainly capable of doing this.

Some amateurs observe at radio frequencies. The Sun and Jupiter are both emitters of radio waves. This activity is more like listening than looking, and is a specialist subject outside the scope of this book. Instead, we will concentrate on observing in or near the visible range of the electromagnetic spectrum.

How Do Lenses Work?

Figure 3.2 shows the phenomenon of refraction [37]. Wherever there is an interface between two transparent materials or a transparent material and air, light changes direction. In Fig. 3.2, the straw appears to bend at the surface.

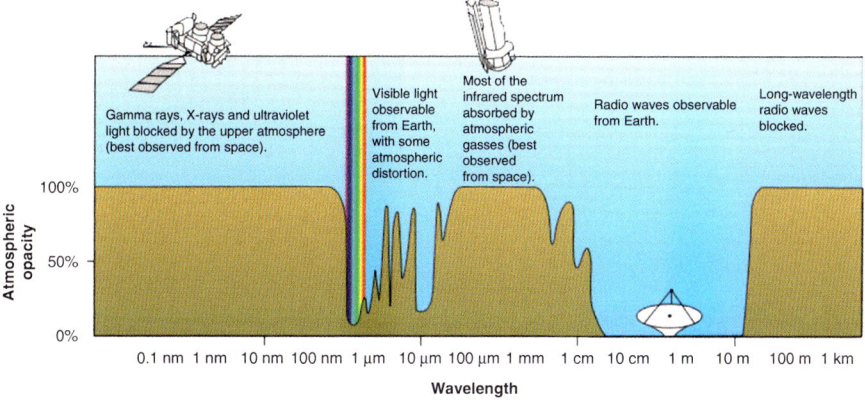

Fig. 3.1 The electromagnetic spectrum. Electromagnetic radiation consists of photons, which have an associated wavelength. Apart from wavelength, all photons appear to be identical (Image courtesy of NASA)

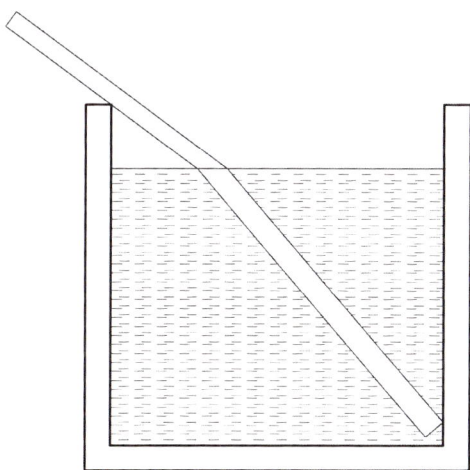

Fig. 3.2 The phenomenon of refraction. Light changes direction at the surfaces and interfaces of materials, leading in this case to the illusion that the straw is bent (Image by the author)

In a lens, we cunningly exploit this phenomenon to bend the light from a given source until it all reaches the same point. This phenomenon is illustrated in Fig. 3.3, although it is only approximately true, as we shall soon see.

A positive or convex lens is carefully designed to refract light so that it is all focused to a point. The distance from the lens to this point is called the 'focal length' of the lens.

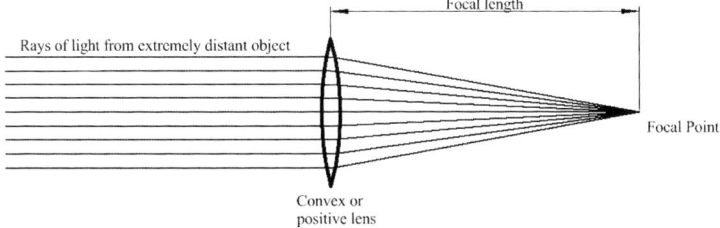

Fig. 3.3 A positive or convex lens is carefully designed to refract light so that it is all focused to a point. The distance from the lens to this point is called the 'focal length' of the lens (Image by the author)

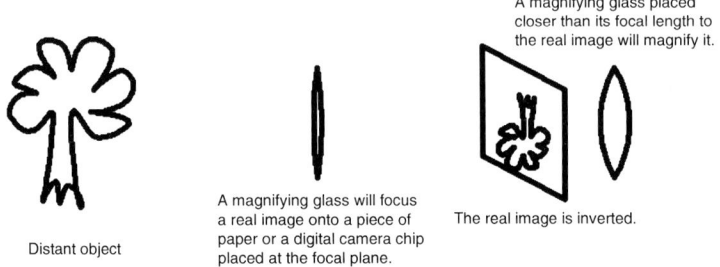

Fig. 3.4 Two magnifying glasses make a telescope! Use the one that magnifies the least to focus an image of an object, as shown here. Use the one that magnifies the most to look at the image you just focused. You can do this with the paper shown, but the effect will also work with no paper there (Image by the author)

You can see this phenomenon for yourself if you take a magnifying glass and hold it a few inches from the wall opposite the window in a room during daylight. A magnifying glass uses a very simple positive lens. An image of the window will appear on the wall. There is a distance from the wall at which the image is sharp. For faraway objects, this distance is the focal length shown in Fig. 3.3. At night, you can often focus a visible image of a light bulb with a magnifying glass. The other thing you can do with a magnifying glass, of course, is to hold it closer to an object than its focal length, look through the glass at the object, and see the object larger than it would otherwise appear.

What would happen if you were to use two magnifying glasses? Figure 3.4 shows. Remarkably, you would have made a telescope. If the two magnifying glasses are identical, you would have a telescope that gives a magnification of 1×, which is not terribly useful. If the two glasses have different focal lengths, the magnification is equal to the ratio of the two focal lengths.

The image quality of what you see, however, is not great. There are two main reasons for this.

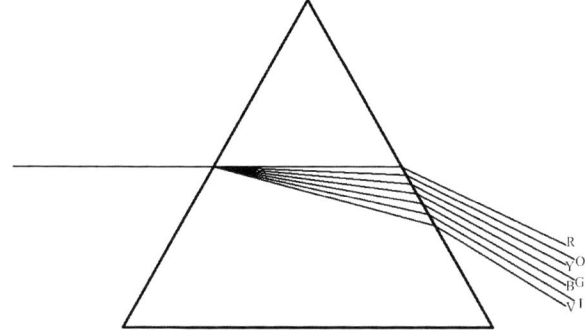

Fig. 3.5 Newton's prism experiment. He showed how white light can be dispersed into the colors of the rainbow, thereby demonstrating that white light contains all these colors. This is because glass reflects different colors by different amounts (Image by the author)

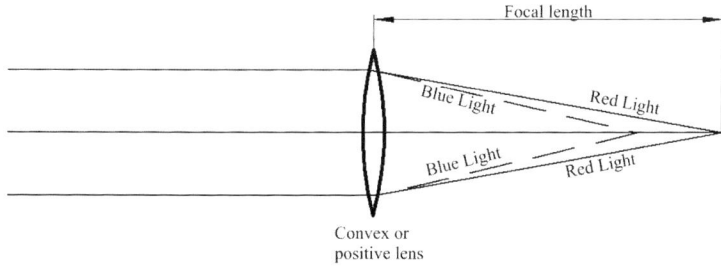

Fig. 3.6 A simple positive lens will refract blue light more strongly than red light. Its focal length for blue light is slightly less than that for red light. This phenomenon is called chromatic aberration (Image by the author)

The first is chromatic aberration. To understand this phenomenon, remember the famous experiment of Isaac Newton with a prism, shown in Fig. 3.5.

Newton showed how white light can be dispersed into the colors of the rainbow, thereby demonstrating that white light contains all these colors. This is because glass reflects different colors by different amounts. There is nothing special about glass in this regard – all transparent materials do it. Just like the prism, a positive lens will refract blue light more than red light. Its focal length for blue light will be slightly less than that for red light. Figure 3.6 shows this.

Incidentally, we can see by comparing Figs. 3.5 and 3.6 that a convex lens can be thought of as a curved prism.

The eyepiece of a telescope is also subject to chromatic aberration. The combined result of chromatic aberration in these two lenses is that objects viewed through the telescope have colored fringes.

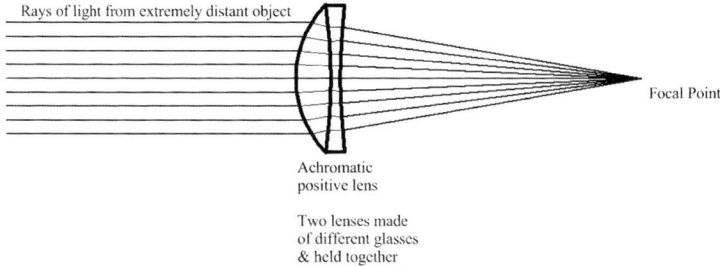

Fig. 3.7 A strong positive lens made of one type of glass can be combined with a weak negative lens made of another type of glass to cause the chromatic aberration between red and blue light to cancel each other out (Image by the author)

This problem was enough of an irritant that some cunning soul called Chester More Hall invented a lens that somewhat corrects for it. His idea was to combine two lenses: a strong positive lens and a weak negative lens, made of two different grades of glass, each with different refractive indices, such that the lens is positive overall. This lens will actually only exactly correct one pair of colors, but it minimizes the effect of other colors sufficiently to make observing life tolerable. Such a lens is called *achromatic*. The principle is shown in Fig. 3.7.

This correction is clearly not perfect. The next level of correction is to join three types of glass together. Such a lens is called *apochromatic*. For achromats, the two glasses used are usually crown glass and flint glass. These are sufficiently inexpensive and easy-to-work with materials that would cause nobody any great pain. The third glass required to correct at three wavelengths is usually fluorite glass, which is expensive and difficult to work with. Apochromatic lenses are therefore expensive, but they do reduce the fringing considerably.

The other main form of aberration is *spherical aberration*. Figure 3.3 would really look more like Fig. 3.8 for monochromatic light.

Concavo-convex lenses, like the one shown in Fig. 3.9, have less spherical aberration than bi-convex lenses.

Interestingly, our eyes correct for both chromatic and spherical aberration. It is curious that such a wonderful optical instrument as the human eye has all these corrections, but is still prone to being short sighted, far sighted and astigmatic (slightly cylindrical lenses). This is at least partly because our eyes evolved to last for much shorter life spans than we actually experience. Most apes don't live more than about 40 years. Before the Industrial Age, we usually didn't either.

It rather seems as if short sightedness is a response to spending too much time on close-up work. It is scandalous that our schools and others do so little to help us take better care of our eyes.

We now know enough to look at refracting telescopes for astronomy (Fig. 3.10). The bare essentials are an objective lens, an eyepiece, a focusing mechanism, and a means to attach it to a mount (not shown). A means to protect the objective lens from dewfall is needed in all but the driest of climates. The most common is to

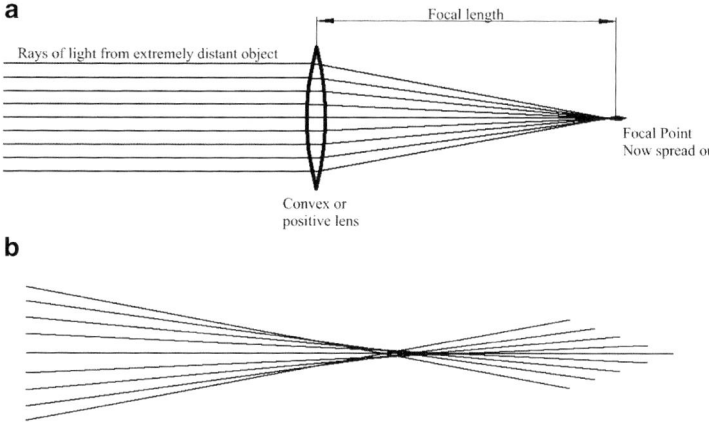

Fig. 3.8 (**a**) The outer parts of a bi-spherical positive lens have a shorter focal length than the inner parts. The lower image (**b**) shows a close up of the region where the rays converge, or, rather, fail to converge (Image by the author)

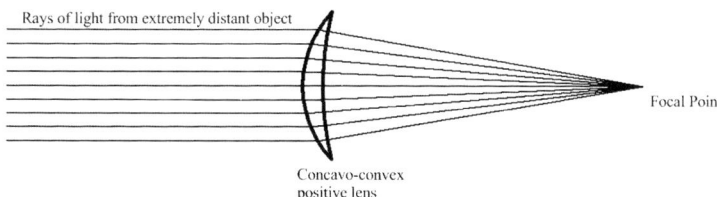

Fig. 3.9 A concavo-convex lens partly corrects for spherical aberration (Image by the author)

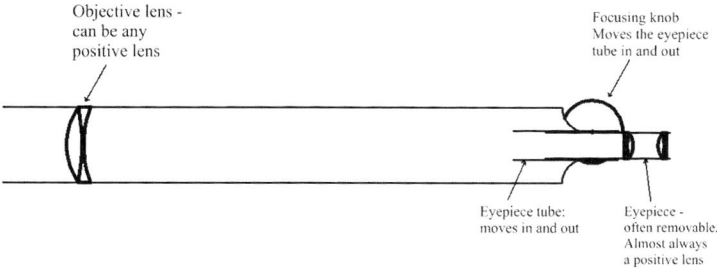

Fig. 3.10 An astronomical refracting telescope. The bare essentials are an objective lens, an eyepiece, a focusing mechanism, and a means to attach it to a mount (not shown) (Image by the author)

extend the telescope tube well beyond the lens, although many manufacturers skimp too much on dew hoods. You can extend them by wrapping black electrical insulating tape around a large plastic cola bottle with the neck and base cut off to leave the parallel section. Cut down the length of the tube and wrap it a little more to decrease the diameter. You can also combine bottle carcasses to increase the diameter. (This won't work too well for a 10" diameter 'scope; a little more imagination may be needed here.)

The telescope tube is not absolutely essential. It is more of a convenience. The insides of the tubes really do need to be as black as soot to keep reflected light to a minimum, especially on moonlit nights. Frankly, too often they aren't black enough. On the other hand, the tubes keep dew out. This is a classic case of an engineering compromise. No one design solves all the problems.

A significant disadvantage of refracting telescopes is that you have to jack them up really high to look through the eyepieces comfortably. The most common work-around for portable telescopes is to have a removable 45° mirror in front of the eyepiece. The price paid is that the image is inverted in one plane.

How Mirrors Work

Any light incident on a mirror is reflected at the same angle (Fig. 3.11). An optical mirror exploits this principle to bring all the light from a distant source to a common focus (Fig. 3.12).

To focus light that is exactly on-axis requires a paraboloid. Most good amateur telescopes' main mirrors (known as *objectives*) are paraboloidal. Cheaper ones sometimes have spherical lenses.

Off-axis, you get a better result with a hyperboloidal mirror. Unfortunately, these are difficult to make. At the time of this writing, they are becoming available on a few high-end telescopes, but generally speaking you get a paraboloidal or spherical mirror. You only really see an advantage if you want to try wide-angle imaging of deep space objects such as nebulae.

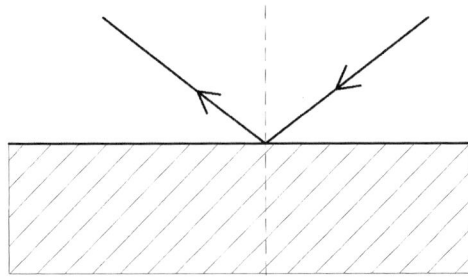

Fig. 3.11 Any light incident on a mirror is reflected at the same angle (Image by the author)

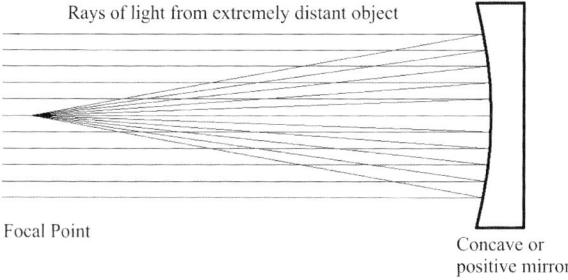

Rays of light from extremely distant object

Focal Point

Concave or
positive mirror

Fig. 3.12 A concave mirror is designed to focus light to a point (Image by the author)

Mirrors have two great advantages over lenses for objectives. First, they do not suffer from chromatic aberration. Second, they are cheaper to make with large diameters. You get a much larger mirror for your money.

Reflecting Telescopes

Although several designs of reflecting telescopes emerged, the one that came to dominate the amateur market was the first, the Newtonian. This is partly for reasons of simplicity. Its manufacture only requires one curved mirror. It is also partly because of its convenience. The eyepiece is to one side, which saves you putting your back out every time you try to look through it.

The principle of the telescope is shown in Fig. 3.13.

Light enters the telescope through an open end and travels to a concave (positive) mirror at the base. It is then focused back. A small secondary mirror reflects the still-converging light to one side, where it is brought to a focus. An adjustable focuser sits at the focal plane. The focuser can hold an eyepiece or a camera, just as with a reflecting telescope.

Some light is lost to the telescope because it is blocked by the secondary mirror and its holder, which is often called a 'spider,' because it is made as thin as possible to reduce light loss and such aberrations as diffraction. A disadvantage of this arrangement is that the spider tends to be easily knocked out of alignment, and you have to keep re-collimating Newtonian telescopes.

The first time you ever collimate a telescope will be a rather daunting experience, but with practice it only takes a couple of minutes. The first thing to note is that even though you can adjust the primary (objective) or secondary mirror, it is usually the secondary that needs adjusting.

The simplest device for this is called a Cheshire collimator. It mostly consists of a pair of cross-hairs in a metal holder, which you insert into the focuser instead of an eyepiece. There's a bit more to it than that because it also has to let some light into the telescope. The idea is that you see two images in the cross-hairs, the original and a reflected image. You simply adjust the secondary mirror until they coincide.

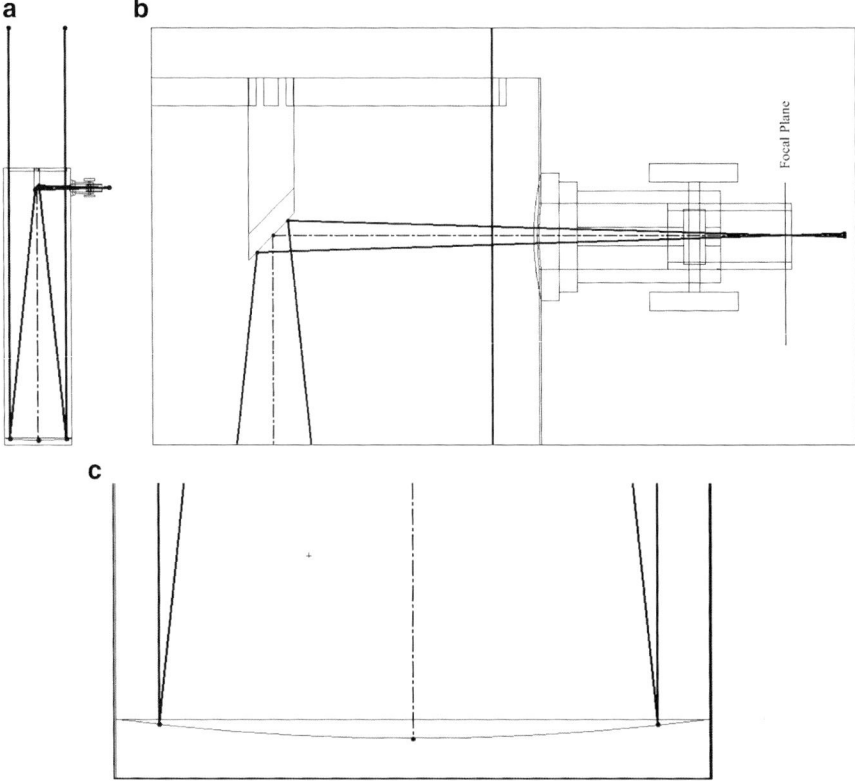

Fig. 3.13 The principle of the Newtonian telescope. Image (**a**) shows the whole telescope. Image (**b**) shows the objective mirror at the base of the telescope. Image (**c**) shows a close-up of the secondary mirror and the focal plane. The focuser can be used to hold an eyepiece or a camera (Image by the author)

It is much easier to use a Cheshire collimator in daylight. You can also buy laser collimators, which work better than Cheshire collimators in the dark. In theory, you probably want to do your collimating in the dark. Some people do this every time they set up. Others rarely do it. If the telescope rarely travels, once every couple of weeks is sufficient. A car ride is usually enough to knock a Newtonian out of collimation.

The second form of maintenance is that the mirrors need to be re-silvered from time to time. There are firms that offer this service. The primary mirror can usually be unscrewed. You mail it off, and a few weeks later it comes back with shiny new silvering. In principle you need to do this annually. In practice, nowadays, silvering firms also offer a thin transparent coating on top of the metal, which will last for years and protect the metal of the mirror from tarnishing.

You can also remove the primary mirror to clean it. The main thing here is to be careful. Mineral particles can and do scratch. You can use tap water and that miracle of modern technology known as washing up liquid on one side of the Atlantic and dishwashing soap on the other. If your tap water is hard, you may wish to rinse afterwards with distilled water. It is as easy to remove the dirt with running water and your hands as with a lens cloth, and you can feel the dirt and be careful with it. Above all, don't be an over-enthusiastic mirror cleaner. The silvering is fragile.

Because mirrors cost less than lenses, it is tempting to buy the biggest darned light bucket you can get your hands on. There is something to be said for this. The resolving power of wide aperture lenses is higher than that of narrow ones. Therefore you can see more stuff. There are downsides, though.

For one thing, big telescopes are unwieldy. They need much stronger mounts to hold them, and if you do not have a permanent observatory, they are difficult to set up. An 8-in. telescope is close to the biggest telescope you want to carry from your garage to your mount. Carrying a 12-in. telescope is likely to be a two-person job.

For another thing, much above 8 in., and the atmosphere can be a limiting optical feature. On many nights in populated areas, there is so much atmospheric turbulence that you will not see clearly through any telescope. If you own a telescope larger than 8-in. diameter, you can expect that there will be nights when the advantage of the larger telescope simply shimmers away.

Cue the *catadioptric* telescope.

Catadioptric Telescopes

There are two main things to know about a catadioptric telescope. First, the light follows a folded path, which makes the telescope a lot shorter than an equivalent Newtonian. Second, it has an objective lens, called a corrector, at the front of the telescope. The secondary mirror is fixed onto the back of this lens, so that the 'spider' arrangement of a Newtonian is unnecessary.

Schmidt-Cassegrain Telescopes

Figure 3.14 shows schematically the layout of the objective system in a type of catadioptric telescope called a Schmidt-Cassegrain telescope, or SCT. Figure 3.15 shows an actual SCT.

The mirrors tend to be spherical, not paraboloidal, which cuts manufacturing costs. The primary function of the corrector plate is to correct for spherical aberration. It does this by being a positive lens towards the central axis but a negative lens at its periphery. That way, the light arriving at the middle of the telescope is forced to focus closer to the average position of focus, while that arriving at the periphery of the telescope is forced to focus further away, thereby undoing the effect of spherical aberration.

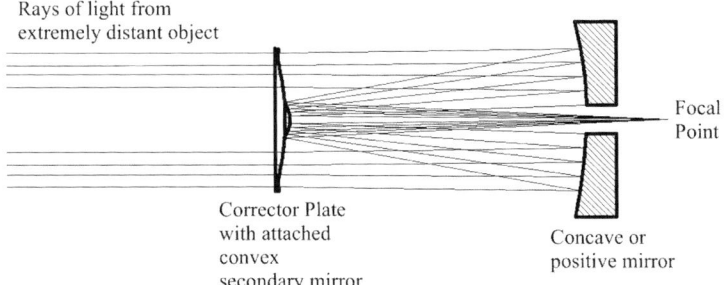

Rays of light from
extremely distant object

Focal
Point

Corrector Plate
with attached
convex
secondary mirror

Concave or
positive mirror

Fig. 3.14 The objective system in a catadioptric telescope. This consists of a corrector plate, which is a glass lens, a primary concave or positive mirror, and a secondary convex, or negative mirror. An eyepiece or a camera may be placed beyond the focal point as with other telescopes. This particular design is a Schmidt-Cassegrain telescope (Image by the author)

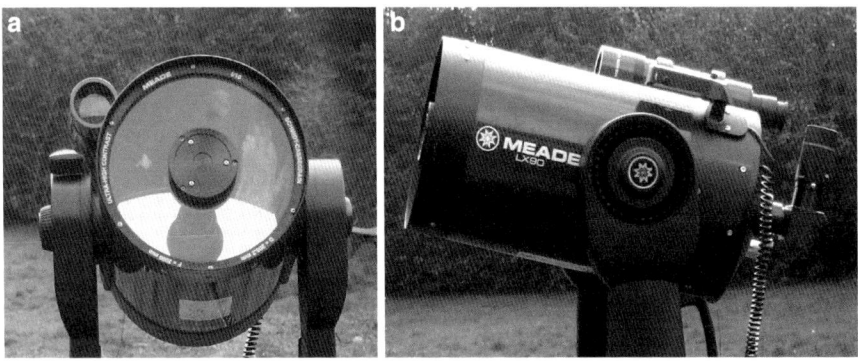

Fig. 3.15 (**a–b**) A Schmidt-Cassegrain telescope in practice. Notice how compact this 8-in. telescope is (Images by the author)

This corrector plate introduces a small amount of chromatic aberration. In practice this is barely noticeable.

Professional telescopes often employ hyperboloidal mirrors because you get less optical aberration off-axis. If the mirrors are hyperboloidal, the telescope is known as a Ritchey-Chrétien telescope. The advantage to this is that a form of aberration called coma, in which point objects such as stars look like little badminton shuttlecocks when off the optical axis, is eliminated. The disadvantage is that hyperboloidal lenses are fearsomely expensive to make.

At the time of this writing, a few higher priced amateur telescopes are becoming available with so-called advanced coma-free (ACF) optics. They mutter the hallowed words Ritchey-Chrétien and remind us that the Hubble telescope was made that way. The truth is more prosaic. The primary mirror in ACF telescopes is still

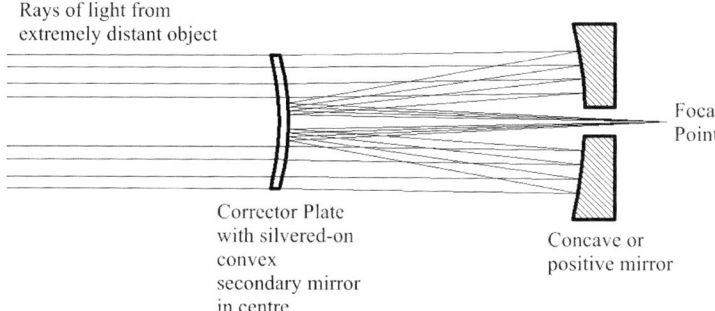

Rays of light from extremely distant object

Focal Point

Corrector Plate with silvered-on convex secondary mirror in centre

Concave or positive mirror

Fig. 3.16 The objective system in a Maksutov telescope. The corrector plate is now a thick negative meniscus glass lens, a primary concave or positive mirror, and a secondary convex, or negative mirror. An eyepiece or a camera may be placed beyond the focal point, as with other telescopes (Image by the author)

spherical, and unlike true Ritchey-Chrétien telescopes, these instruments require a glass corrector plate. Careful reading of the promotional advertising suggests that the secondary mirror is the only hyperboloidal one, and that the manufacturer has redesigned the corrector plate to reduce coma.

Ritchey-Chrétien telescopes actually suffer somewhat from another aberration, astigmatism. This occurs off-axis when the image passes through an optical path where the curvatures are slightly different in mutually perpendicular directions. In practice professional astronomers live with this because it is relatively easy to allow for in astrometric calculations.

SCT telescopes focus differently from Newtonians and refractors. The eyepiece is not moved relative to the main tube of the telescope. Instead, the primary mirror is wound back and forth by turning a knob on the back of the telescope. In general this works well because the telescope is so compact that the mount is not greatly disturbed by manual focusing.

There is usually no capability to re-align the primary mirror in SCTs. This is not a serious problem. You can and should collimate the secondary mirror. For this, an accessory called Bob's Knobs is highly recommended. By replacing the collimation screws with them, you can collimate without a screwdriver, and thereby greatly reduce the risk of accidentally scratching the corrector plate.

Maksutov Telescopes

This type of telescope is a variant on the catadioptric theme. It has a different corrector plate, which is so curved that the secondary mirror can be painted or vacuum deposited straight onto it. Figure 3.16 shows a Maksutov telescope.

Maksutov invented his telescope with two objectives – to eliminate the need to collimate and to seal the mirrors in so that they would never need re-coating. (Presumably this means that there has to be a flat glass plate in front of the eyepiece.)

In these aims, Maksutov succeeded, but as with all engineering solutions, he had to compromise elsewhere. The first disadvantage of the Maksutov is actually an advantage to Solar System observers. The optics only work well if the focal length is long, so the field of view is narrow. In that event they work very well. There is little chromatic aberration or coma. This is perfect for observing planets, because the high magnification of a long focal length wins against the narrow field of view. Through an f/5 or f/6 1,200-mm focal length telescope you see an awful lot of blank sky even around Jupiter. A Makustov may have a focal length of 2,700 mm. This is not the focal length you would want to view, say, the great nebula M42 in Orion, but it is great for planets.

The second disadvantage is that the corrector plate is heavy. This means it can take a long time – 2 h is usually quoted – to cool down in the evenings if it is kept indoors. It also means that you need a better mount to guide it.

The cost of making the Maksutov corrector plate means that 7 in. is the maximum practical diameter. This kind of telescope survives in the market because it is good for observing planets.

Eyepieces

Some people collect eyepieces like others collect stamps or coins. There are definitely good and not so good eyepieces.

The magnification you obtain from your telescope is simply the focal length of the objective divided by the focal length of the eyepiece. Thus a 1,200-mm focal length telescope would get 48× magnification with a 25-mm eyepiece, and 84× with a 12.5-mm eyepiece. Most people advise against expecting too much in the way of magnification. There comes a point at which you mostly end up magnifying the atmospheric turbulence; you can actually see better with modest magnification. If you really need more magnification, buy a bigger telescope with a wider objective lens and/or a longer focal length.

The number of glass elements in the eyepiece is not the most exciting issue here, although manufacturers promote this heavily. You can certainly buy six-element eyepieces, and they are usually better than four-element eyepieces. If your telescope will accept 2-in. eyepieces you will get a better quality image than with 1.25-in. eyepieces, but the 2-in. ones – and filters for them – cost a lot more. You often also get better quality glass and anti-reflective coatings if you pay more.

Three other issues are at least as important, possibly more so: chromatic aberration, eye relief and confocality.

Eyepieces do suffer from chromatic aberration. If your budget stretches to the more expensive, apochromatic eyepieces are definitely worth considering.

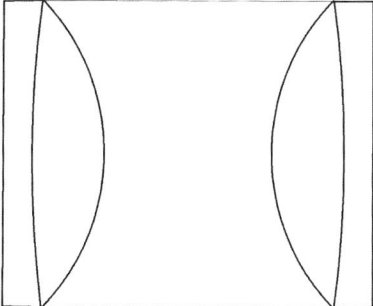

Fig. 3.17 The Plossl or Plössl eyepiece is an achromat with four lens elements (Image by the author)

Confocality is not a property of a single eyepiece but of a range of eyepieces. What it means is that if you use a low-magnification lens to find your object, and then want to put in a high-magnification eyepiece to see it better, if the eyepieces are confocal, then you don't have to re-focus the telescope. That's an unadulterated plus, apart from the money. In practice this tends to mean buying your eyepieces as a set.

Before getting onto eye relief, we will look at the most popular eyepiece, a design called a Plossl or Plössl. These eyepieces are the staple of the low-cost eyepiece market.

Figure 3.17 shows a schematic view of this four-element design, which is quite prone to internal reflection and quite demanding of glass quality. So you get a better eyepiece if it has good anti-reflective coating and good glass. You can also get variants called Super-Plossls, which usually have the same lenses but better coatings and glass.

The trouble with these eyepieces is that you have to get your eyeball very close to them. This is useless if you are an eyeglass wearer. You yourself might not be an eyeglass wearer, but you can bet that a good few of the friends and family you wish to impress with your telescope are. If the eyeglass wearer has no astigmatism, the telescope can be re-focused for them, but this is a nuisance, and if showing your friends, you have no quality control over their focusing, as non-experts have difficulty focusing well.

The distance you can put between your eyeball and the eyepiece is called the available *eye relief*. You want at least 20 mm, or 0.8 in., of eye relief for an eyeglass wearer. The thing to look for when buying an eyepiece is a *long eye relief* eyepiece if you expect an eyeglass wearer to use it.

Finally, if you have paid big bucks for a good telescope, why ruin the view with a cheap and nasty eyepiece (Fig. 3.18)?

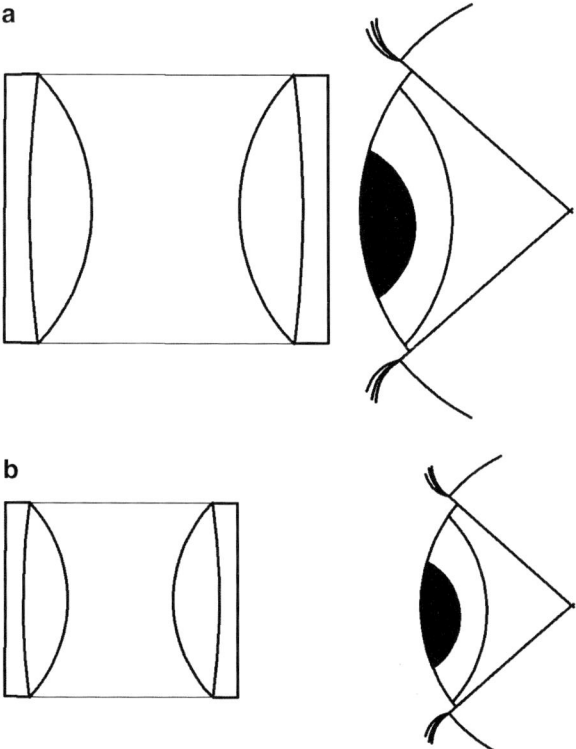

Fig. 3.18 A short eye relief eyepiece (**a**) and a long eye relief eyepiece (**b**). For eyeglass wearers, long eye relief is better (Image by the author)

Barlow Lenses

A Barlow lens is simply a negative lens that you can plug into your eyepiece to increase the magnification (see Fig. 3.19.) You can also use such a lens with a camera to obtain the same benefit.

The same remarks apply to Barlows as to eyepieces – the better the glass and coating, the better the Barlow lens. Apochromats are better than achromat, not least for photography. You should be able to keep the light loss down to 3 % with a good Barlow.

So is a Barlow-plus-short-focal-length-eyepiece better than a short-focal-length eyepiece? All other things being equal the answer is likely to be yes, because the short focal length eyepiece has more curvature and therefore more aberration. And, you get better eye relief.

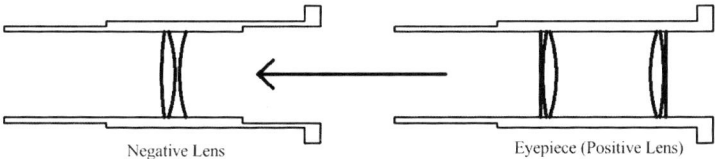

Negative Lens   Eyepiece (Positive Lens)

Fig. 3.19 A Barlow lens is a negative lens you can insert in front of your eyepiece or camera to increase the magnification (Image by the author)

The combination of a Barlow lens and an eyepiece is actually a zoom eyepiece. By varying the distance between the eyepiece and the Barlow lens, you can change the magnification you achieve. Therefore you should treat the claimed magnification boost with a pinch of salt. Clark [38] reports that with a webcam, a particular 2× gives 2.3×, and a particular 4× Barlow gives 3.4×.

The Pros and Cons of Various Telescopes

You can draw up your own table of comparisons, like Table 3.1.

You can use this table to help choose your telescope type. Bear in mind, however, that most of these pros and cons are pretty marginal for Solar System objects. It does not much matter which type of telescope you choose. What matters is its size and quality. It is better to spend your money on a really good mount than the latest, greatest optics.

Dobsonians are much more portable. You can carry a 12-in. or even a 16-in. around if you are strong. The 18-in. Dobsonian owned by Bristol Astronomical Society in England has just been installed permanently because it proved to be too heavy to carry around.

Binoculars

You are likely to be disappointed unless you can mount your binoculars on a tripod. However, most pairs of binoculars do not come with a means to mount them to a tripod. You can buy adaptors to mount them onto a camera tripod, or, of course, if you are reasonably handy, you can make an adaptor.

There is no reason not to mount the binoculars onto a motor-driven mount that will track the heavens, although people rarely do.

Camera tripods are perfectly good for this purpose. Even a monopod is better than nothing, although it is not as good as a tripod. You will see better if the binoculars are not picking up your bodily tremors. The advice given to photographers is that cheap, flimsy tripods are inadequate. The same applies when they are used with

Table 3.1 A comparison of telescopes

Telescope type	Advantages	Disadvantages
Refractor	Good contrast because there are no mirrors to tarnish	Very long
		Large models are expensive
	No need to collimate	Chromatic aberration is expensive to correct
		Needs a deep lens hood to prevent misting due to dew-fall
Newtonian reflector	Large primary mirrors are relatively inexpensive	Mirrors need periodic re-silvering
		Needs regular collimation
	Not very prone to the effects of dew-fall	High-diameter models can become long
Schmidt Cassegrain	Compact	Needs a deep lens hood to prevent misting due to dew-fall (Commercial dew hoods are not very impressive.)
	long focal lengths mean high magnifications	Care needed not to scratch the corrector plate
	Needs no re-silvering	Needs regular collimation
Maksutov	Compact	Heavy corrector plate slow to cool down
	Very long focal lengths mean high magnifications	Heavy corrector plate means strong mount needed
	Needs no collimation or re-silvering	Not available above 7-in. diameter
	Good optics at high magnifications	

binoculars. If new ones are too much money, you can of course find used ones online or wherever if you would rather spend less. You will get more use out of a battered, good quality tripod than you ever will from a shiny new one that shakes like a leaf. There seems to be a bottomless market for tripods.

What should you know about buying binoculars for astronomy?

The first thing to know is that the main market for binoculars is bird watching. Engineers tend to design binoculars to be just about adequate for this purpose. This is no crime. They have to sell into a competitive market and can't afford to throw money around on over-engineered products. The thing is to realize that binoculars for this purpose are not really designed for astronomy.

The lower-cost ones tend to be used by casual hikers who only practice their hobby in daylight. The primary purpose of the binoculars is to magnify, not to gather light from objects so dim that they can only be seen at night. Hikers are also likely to place a premium on compactness and light weight. There has been a tendency in recent times to give the lenses of bird-watching binoculars a red tint. Indeed it is getting to the point where most binoculars are tinted red.

For astronomy, on the other hand, you want binoculars that gather as much light as possible and deliver it into your eyes. Magnification is important, but it can be overrated. You also do not really want to throw away blue light by making your lenses into red filters.

The way binoculars are sized is by numbers such as 10×50 or 7×30. The first number is the magnification. The second number is the diameter in mm of the objective lenses (the big lenses at the front). Thus a pair of 7×50 binoculars has 50-mm diameter objective lenses and magnifies about seven times. They have almost 2.8 times the light-gathering power of 7×30 binoculars because that is how much bigger the area of the objective lenses is. (Area is proportional to diameter squared.)

All eyepieces, binocular or telescopic, exhibit the phenomenon of eye relief. Most binoculars are not very eyeglass-wearer friendly.

Another trap for the unwary binocular buyer is that the light exiting the eyepiece is sometimes concentrated into too narrow a beam to use the whole of your open eye pupil. This wastes eye capacity at night when the pupils of your eyes should be fully dilated. For daytime bird watching it does not matter: the binoculars probably then gather more than enough light, and the user's pupils are likely to be quite tightly closed.

Binoculars can usually be adjusted to compensate for different people's eye separation, though not every user does this. It is worth doing. For terrestrial use, binoculars not only magnify the view, but also allow the user to see the object in stereo. The fact that the objective lenses are further apart than the user's eyes increases the stereo effect.

Some people do not have stereo vision. If that applies to you, you will not experience this particular benefit of binoculars.

Of course, you cannot see celestial objects in stereo. Even the International Space Station is too far away to see in stereo. The difference between the two images is smaller than the pixel size in your eye, and so cannot be resolved. Yet curiously seeing objects with two eyes can make them appear more three-dimensional. You can see this effect by looking at the Moon through binoculars. There is something that your brain does if you see an object through both eyes. An impressive example of this was demonstrated at a star party, where someone was letting people look through an unusual pair of binoculars that he had made. These binoculars consisted of two 12-in. reflecting telescopes, with prisms so arranged that you looked through one eyepiece with each eye. This man was showing the globular cluster M13. It really did look spherical. This cannot have been due to stereoscopic effects – M13 is over 20,000 light-years away. It must have been an illusion of mental interpretation. If you are fortunate enough to have two good eyes, there is something very reassuring about seeing something with both eyes that a telescope simply does not afford you.

To work well, the two halves of the pair of binoculars must be as perfectly parallel as possible. The jargon for adjusting the parallel effect is *collimation*. Most binoculars are collimated once in the factory and cannot be re-collimated. Since they are glued together, they cannot even be stripped down and reassembled.

West Norfolk Astronomy Society once acquired a pair of 20×80 binoculars, which quickly went out of collimation and showed double images of everything. They were not cheap, and everyone was very annoyed. You have to spend lots of money to get binoculars that can be re-collimated. (Such binoculars are also heavy

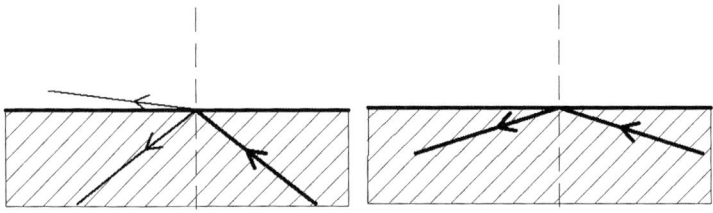

Fig. 3.20 Internal reflection. When a ray of light leaves a piece of glass, two things happen. Some of the light is reflected back through the glass. The rest leaves it, but it is refracted, and it leaves the glass at a more glancing angle than that of the incident ray (*left*). If that glancing exit angle is zero or less, no light leaves the glass. It is all reflected (*right*) (Image: Author)

and need a heavy-duty tripod.) The store offered replacements, but they actually had worse collimation than the original. Good astronomical binoculars ought to be designed either to withstand a 6-ft fall onto grass, or be re-collimatable afterwards. Let the buyer beware!

A pair of 20×80 binoculars will show Jupiter's moons quite well, and you can just about resolve the largest bands on that planet. They will not, however, resolve Jupiter's innermost moon Io coming out of eclipse. Through them, you can see the phases of Venus and can just about see Saturn's rings.

Binoculars are of course two telescopes joined together, one for each eye. They have one more optical trick up their sleeve, however. The image you see is upright, not inverted.

This is done via a double Porro prism. To know how this works, we need another optical concept, total internal reflection. Figure 3.20 illustrates this. When a ray of light leaves a piece of glass, two things happen. Some of the light is reflected back through the glass. The rest leaves it, but it is refracted and leaves the glass at a more glancing angle than that of the incident ray. As the incident angle moves away from perpendicular, the exit ray leaves at a more and more glancing angle. Eventually the exit angle is zero, but the incident light angle is not. If the incident light is more glancing than this, the light has no choice but to be reflected. You then have total internal reflection.

This principle is exploited in Porro prisms (Fig. 3.21). In cross-section this is a right-angle triangle. Light perpendicularly incident on the hypotenuse side is not refracted and so suffers no chromatic aberration. This is because the incident angle is 90°.

When the light reaches the next face of the prism it undergoes total internal reflection, also a chromatic-aberration-free process. It experiences a second total internal reflection at the last face of the prism, before exiting perpendicularly through the face by which it enters. Again, therefore, there is no refraction. There is no chromatic aberration. This is because the incident angle is 90°.

When a virtual image leaves a Porro prism, it leaves as a mirror image of the image that entered (Fig. 3.22).

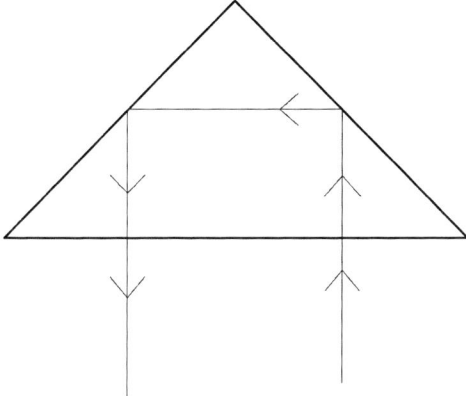

Fig. 3.21 A Porro prism (Image by the author)

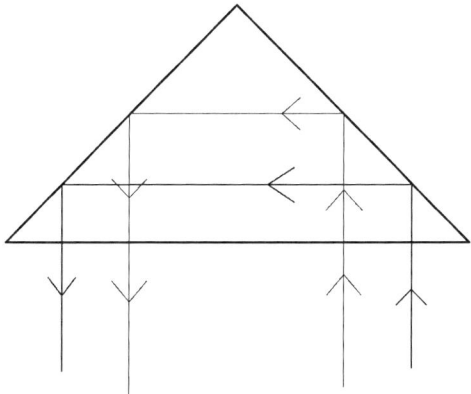

Fig. 3.22 Two rays entering a Porro prism leave it on opposite sides of each other (Image by the author)

These facts can all be exploited by placing two Porro prisms at right angles to one another as in Fig. 3.23. By tracing the rays of the letter F you can convince yourself that these two Porro prisms will convert an inverted image into an erect one.

The other thing the prisms do is to move the image laterally, enabling the eyepieces to be closer together than the objective lenses. This is essential for all but the tiniest objective lenses (Fig. 3.24).

Binoculars are usually achromats, not apochromats. This is very noticeable when you look at the Moon through large binoculars. That said, you can get stunning views of the Moon through 20×80 binoculars, despite the color fringing.

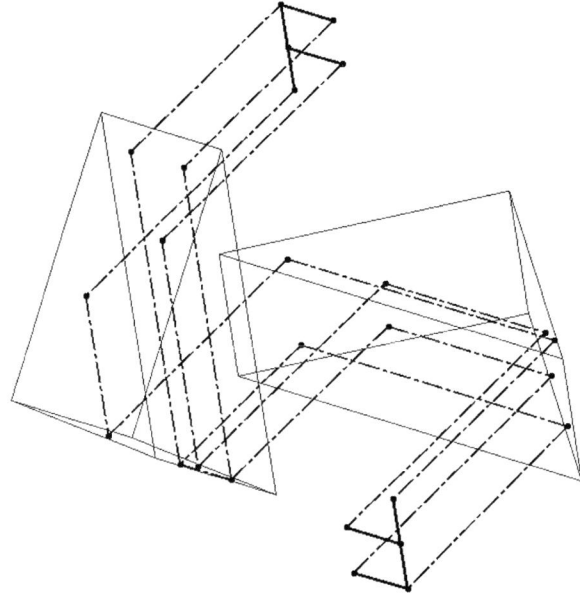

Fig. 3.23 A virtual image of an upside-down letter F is sent through two mutually perpendicular Porro prisms and in the process it is erected (Image by the author)

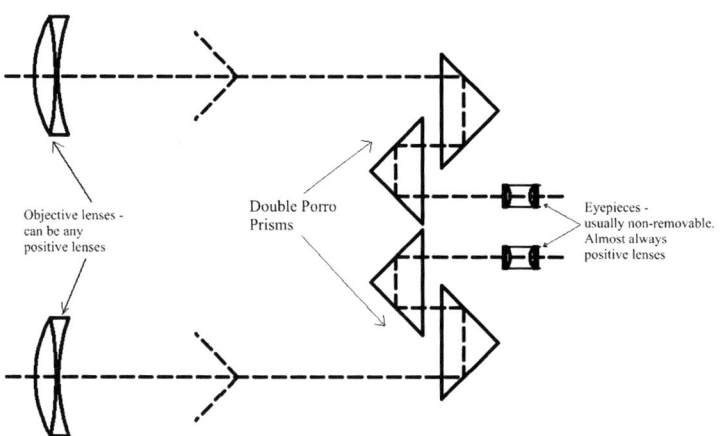

Fig. 3.24 A two-dimensional schematic image showing the optical path in a pair of binoculars. In reality the Porro prisms cannot all be in the plane. Two of them have to be out-of-plane (Image by the author)

Binoculars have a hallowed place in astronomy for helping find celestial objects quickly, in which case they need to be hand-held. For all other jobs, they will give more satisfactory results if mounted on a tripod. Not all binoculars have a thread to be placed on a mount. However, you can buy special-purpose holders.

Filters

The visible part of the electromagnetic spectrum can of course be seen as the colors of the rainbow. Each color corresponds to a particular wavelength of light. Table 3.2 lists the wavelength of each color. The metric units, nanometers, correspond to billionths of a meter. The wavelengths of light are indeed short.

The purpose of filters is to enhance views of celestial objects by selectively removing light at certain wavelengths.

The Moon looks far too bright through a telescope. You really need a neutral density filter once it is properly dark. You can either use a gray filter or a pair of polarizing filters, which you rotate to obtain the desired amount of filtration. This is probably the better solution, as the brightness of the Moon depends on how dark it is and how clear the sky is.

You can get 'fringe killer' filters that remove the purplish halo around objects, by getting rid of purplish and UV light. Indeed infrared/ultraviolet (IR/UV) blockers are essential for digital astrophotography, because the chips are sensitive to more than just visible light. They have the additional advantage that they keep dust off your camera's chip.

Light pollution filters are quite helpful with sodium (orange) street light, especially the so-called didymium filters. Didymium is a mixture of the elements praseodymium and neodymium. Glass impregnated with didymium transmits very little of the orange street light, but transmits most other visible wavelengths. It also enhances the red in autumn leaves. You can therefore buy didymium filters called 'redhancers' for cameras for this very purpose, and use them to help with any astrophotography you may wish to engage in with such cameras.

You can use colored filters for various purposes, listed in Table 3.3.

Table 3.2 Wavelengths corresponding to the colors of the rainbow

Color	Wavelength (nm)	Wavelength (millionths of an inch)
Red	700	27.6
Orange	620	24.4
Yellow	580	22.8
Green	530	20.9
Blue	470	18.5
Violet	420	16.5

Table 3.3 Color filters and their uses

Red	Reduces dusky sky
	Brings out some features on Mars, Jupiter and Saturn
Deep yellow	Brings out some features of Jupiter (especially in polar regions), Saturn and Mars
	Make comet tails more visible
Deep green	Particularly good for the atmospheric bands and the Great Red Spot on Jupiter, because they are darkened relative to other features
	Darkens the sky for daylight observation of Venus
Blue	Enhances contrast of Moon
	Brings out some features of Jupiter and Mars
	Make comet tails more visible
Transmits, not blocks, UV, blocks visible light	Shows up the clouds on Venus. Claims are made that yellow, green and blue filters also do this, but nothing brings these out quite like UV light

Filters come in 1.25-in. and 2-in. sizes, and screw into the base of your eyepiece.

The 1.25-in. filters usually have a 1¼×42 turns-per-inch thread. Filters have a male thread at the back and a female thread at the front so that you can use more than one filter at once, e.g., a green filter plus a UV/IR blocker on a camera.

Not all filters have this thread. It is worth a quick check before you buy.

You may need to re-focus slightly between filter colors if your optical path is partly through glass.

Conclusion

Electromagnetic radiation is our tool for detecting Solar System objects. Visible light is a small part of the electromagnetic spectrum, but it is the most useful part for Solar System astronomy.

Numerous telescope types are available. Quality is more important than type. A good quality version of any telescope or binocular type will serve you well for Solar System astronomy.

Next, we will look at Solar System astrophotography.

Chapter 4

Photographing the Moon and Planets

Introduction

One of the easiest ways to become disappointed is to expect your pocket digital camera, or your phone camera, to give you good pictures of the heavens. You might manage a tolerable picture of the Moon; and the brighter planets are at least recognizable if the camera is pointed at a telescope eyepiece. But they are no better than being barely recognizable. This chapter will explain how quite modest equipment will serve you far better.

For the planets and Moon, which are bright objects no dimmer than eighth magnitude, you only need a little camera – size is not the issue – but you need the right little camera. Ironically the last thing you need to take still pictures of these objects is a still camera. You need a camera that takes movies.

In this chapter you will be shown how to screen these frames and throw out those that are all messed up by atmospheric turbulence, selecting only the good ones. If you superimpose these good frames, you will have a picture broadly equivalent to a time exposure under good 'seeing' conditions.

You will then be shown how to post-process these images to bring out the features in them. You will then see details that were not at all apparent when you looked through your telescope.

There is a book from the 1970s called *Asimov on Astronomy* [39], with a dust jacket photo of Saturn taken through a 60-inch telescope. It is no exaggeration to say that it is no better than what you can obtain today with digital photography and a 12-inch telescope. That's how much digital photography has improved astronomy. We can now take pictures from our backyards that only professionals with better

© Springer Science+Business Media New York 2015
J. Clark, *Viewing and Imaging the Solar System*, The Patrick Moore
Practical Astronomy Series, DOI 10.1007/978-1-4614-5179-2_4

telescopes than most university astronomy departments possess could manage a generation ago. Of course, today's professionals have instruments that could only be dreamed about in the 1970s. Everybody's game has moved on.

There are some good books around, such as Covington [40] and Berry and Burnell [41]. There is a third on webcam photography, by Reeves [42]. Reeves' treatment is more theoretical than the one presented here, which is quite deliberately hands-on.

Hardware

You will need a telescope of reasonable power (at least 4.5 inches and preferably 6 inches) and a motor-driven mount of at least the quality of an EQ5 mount.

You will also need a movie camera with a CCD chip. This could be the Solar System imagers made by several companies, such as Meade™, Celestron™, or Opticstar™. These cameras come with their own software, and provide all the hardware and software you need. They are the easy option.

This does not make them the cheapest option. You can get results just as good with a CCD webcam. But be careful here. Most webcams have CMOS chips, not CCD chips. Webcams are now as cheap as chips (the edible kind).

The first one to become popular for this purpose was the Philips ToUcam™. This was superseded by the Philips SPC900NC™. That too has gone the way of all flesh. Unfortunately, most high-end webcams are now made with CMOS chips, which are no good for astrophotography. Webcams are also increasingly built into computer monitors, reducing demand for separate models.

The next bit of DIY you need to turn a webcam into a viable tool for astrophotography is an adaptor to make it fit into 1¼-inch focusers. You have to pry the lens holder off, unscrew the lens, and screw the specially made adaptor into the same hole. You should first fit a filter over the open end of the adaptor to keep dust off the chip. Once you get dust onto the chip, it is the devil's own job to remove it. It will show up in photos. The filter you need is a UV/IR blocker. The chip is sensitive to these wavelengths, and they will ruin your images.

The same applies to the Meade and Celestron cameras. They, too, come with the chip exposed to the elements. This isn't a good idea. Put a filter on as soon as you can, and leave it there.

Figure 4.1 shows the author's collection of Solar System Imaging cameras.

The Meade, Celestron and Opticstar cameras come with their own software, but the Philips does not. It is not a purpose-built astronomy camera. You can get free software to make avi files at http://www.astrosurf.com/astropc/oldversion/qcam/doc/uk_qcfocus1.html. You can also pay for a software package called K3CCD Tools™, which is very good with Philips webcams.

Fig. 4.1 The author's collection of Solar System imaging cameras. Front: A monochrome Meade DSI II. This was designed as a deep sky imager, but it works just fine for Solar System objects. It has a chip size of 752×582 pixels, giving it a slightly wider angle of view than the 640×480 chip on the color Philips SPC900NC camera (*right*). The lens has been removed from this camera. It has been replaced with a purpose-built 1.25″ adaptor. To the rear left is a more modern monochrome Opticstar PL-131 M. It has a much larger 1,280×1,024 chip. Because it is monochrome it has higher resolution than a color camera (see Fig. 8.46). It also has a much higher frame rate than the 5 fps of which the Philips is capable. This is good for capturing the rotation of Jupiter. The large chip makes it a lot easier to keep targets on the chip, at the cost of creating larger avi files (Image by the author)

Video Capture

As we embark on a journey through the various dark arts of astrophotography, please don't think that you have to master everything before you can start taking pictures. You can get away without dark frames at first, although you will get much better pictures if you use them. If you are afraid to 'start small' and work your way up, you may never get anywhere. It is no great sin to make a few dumb mistakes on the way. The stakes are not high; it is only a hobby. Besides, if at first you omit the 'twiddly bits,' you will eventually know why the more advanced techniques are advantageous, and will understand when you really need them.

The most useful form in which to capture webcam frames is into an avi file. The various software packages that do this are all similar in principle. The software package to be used to give you a strip-cartoon-style lesson in the art of webcam image capture will be *K3CCDTools*™ (Figs. 4.2 and 4.3).

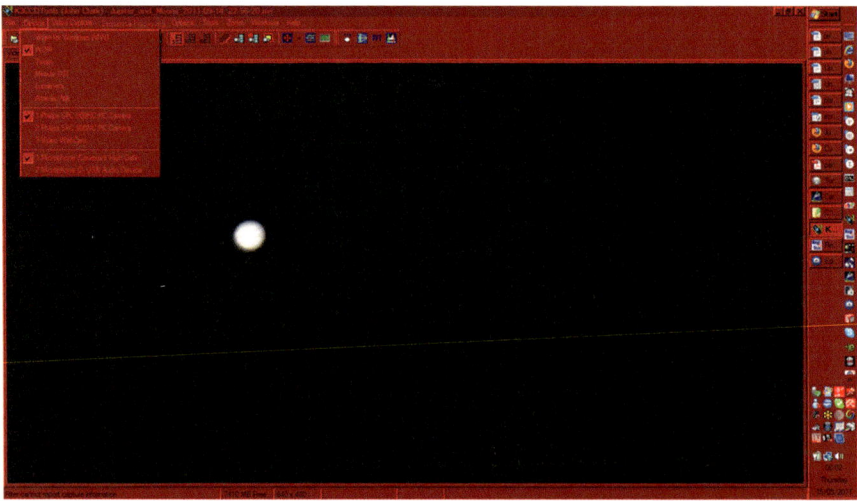

Fig. 4.2 A screen dump from K3CCDTools, showing how your first task is to tell it what kind of camera to look for (Image by the author)

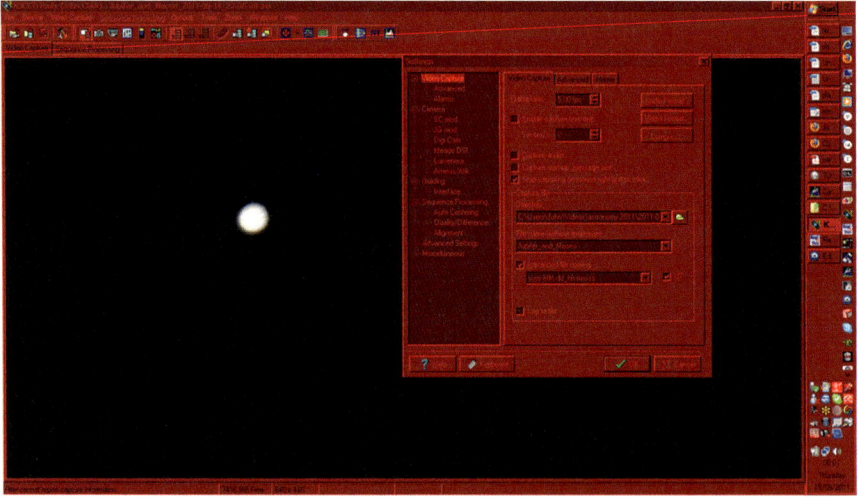

Fig. 4.3 A screen dump from K3CCDTools, showing the settings menu for performing many tasks, the main one of which is to give your avi files names and tell the program where to store them. The feature that puts dates and times into the file names is very useful (Image by the author)

Fig. 4.4 This screen dump from K3CCDTools was taken while recording the webcam image. In this case the program is set to record for 30 s 15 times at 10-min intervals (Image by the author)

Also very useful is the drop-down menu of file names. The one shown is 'Jupiter_and_Moons.' By choosing this naming prefix, together with the 'advanced file naming' feature, you end up with file names like 'Jupiter_and_Moons_2011-09-29_22-44-17.avi.' The program stores naming prefixes. An advantage to using the file names of the form suggested is that you need only a small number of prefixes. This minimizes the amount of typing you have to do outside in the dark (Figs. 4.4, 4.5 and 4.6).

Image Processing

K3CCDTools will do this perfectly well, but we will switch to a free program called *Registax*™ because it is often used with other image capture software besides *K3CCDTools*, such as that supplied with dedicated webcams. You can download this from http://www.astronomie.be/registax. It comes with the Celestron NexImage™ camera.

What follows is a strip-cartoon style tutorial for *Registax* (Fig. 4.7).

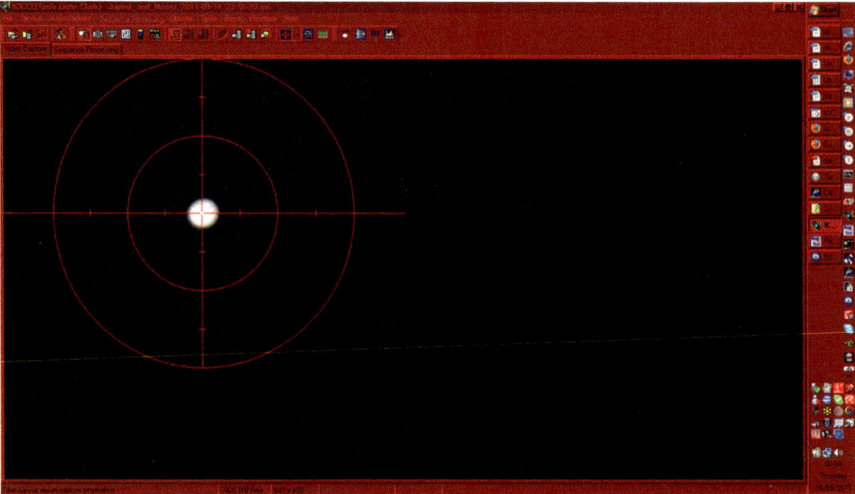

Fig. 4.5 This screen dump from K3CCD tools shows how you can display cross-hairs if you wish to center your object (Image by the author)

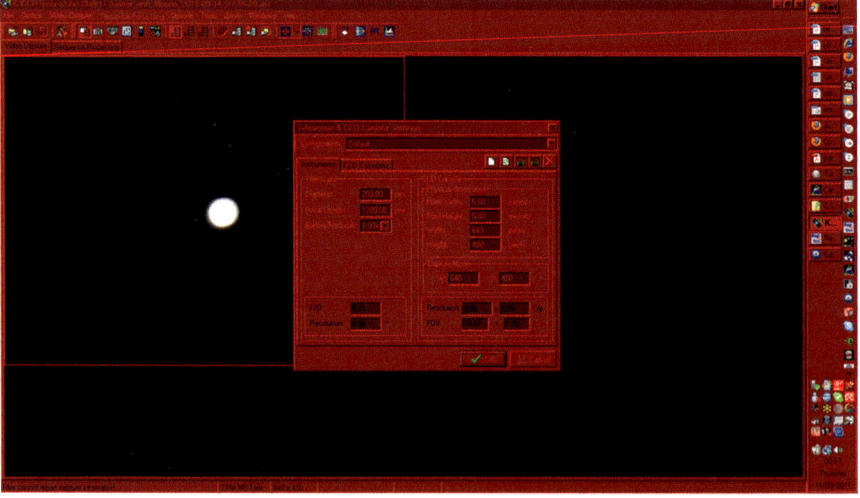

Fig. 4.6 This screen dump from K3CCD tools shows two things. First, you can optionally display the edges of the image. Second, if you enter your telescope's parameters, the program will calculate the size in arcminutes of your field of view, together with an estimate of your resolving power (Image by the author)

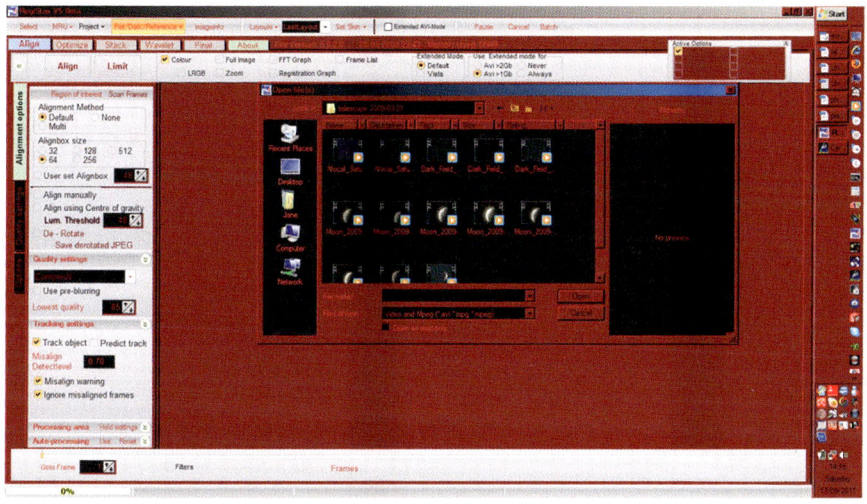

Fig. 4.7 First you select the avi file you want to process (Image by the author)

Now we need a dark frame. What is this? You put the cover on the telescope and shoot about 30 s of footage. This 'image' will be the thermal noise on the CCD chip, which *Registax* will then subtract from actual images that you want to process. A handful of pixels will be 'hot' pixels, which appear very bright no matter what light falls on them. Using a dark field image will get rid of these distracting bright spots.

Using a dark frame image background removes a blue background from your images. This is particularly true if you have the webcam turned to maximum sensitivity to pick up, for example, Saturn's moons, which are faint enough to be at the limit of what a webcam will detect with an 8-inch telescope. The sky will then appear to be almost white even on a cold night.

On hot summer nights to cut down on thermal noise you may wish to wrap your webcam in a thin, flexible plastic film and put it in the refrigerator to cool it down before a night's photography.

The menu shown in Fig. 4.8 can also be used to create a flat frame. In practice this is of limited value with webcams because the image moves around the frame enough to average out the effect of 'dust bunnies' on your chip (Figs. 4.9, 4.10, 4.11, 4.12, 4.13 and 4.14).

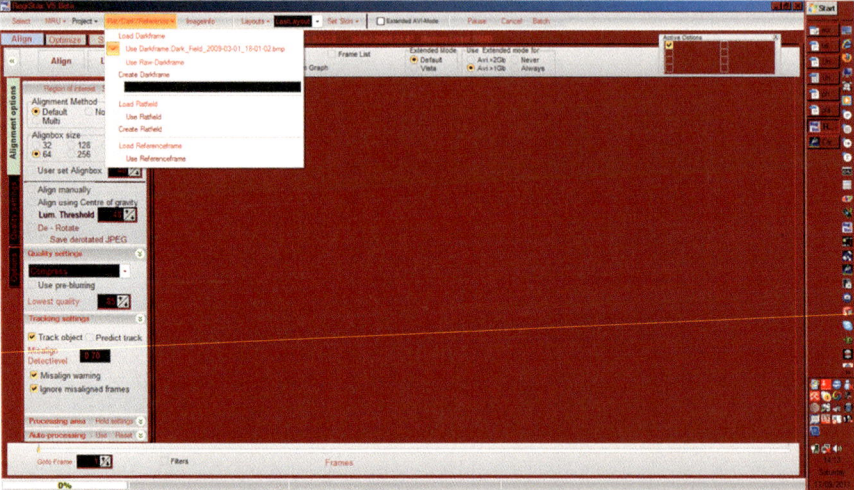

Fig. 4.8 Use this menu to create a dark frame and a flat frame. You select an avi file in the usual way and press 'Create darkframe.' Registax will then process the movie clip and stack the frames to make a dark frame. Having saved it as a bmp file, you load it, and it will thereafter use this as the dark frame until you load another (Image by the author)

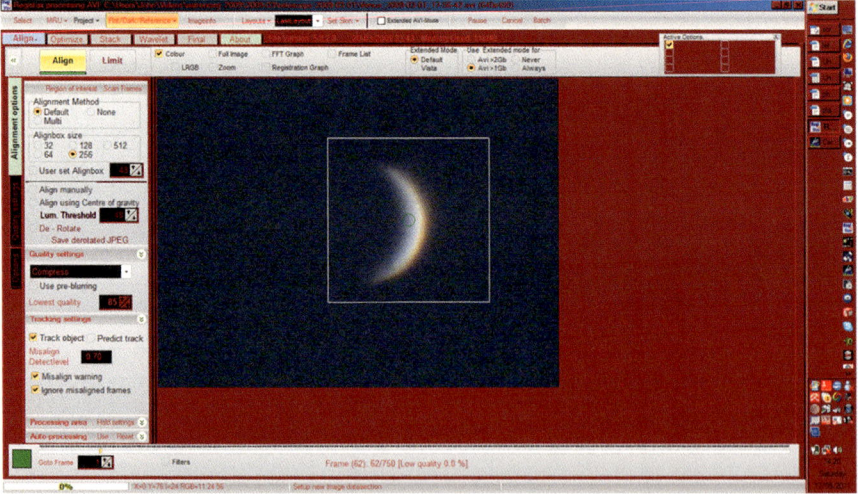

Fig. 4.9 Now you select a good frame by moving the slider along the fame selector bar at the bottom of the Registax window. This is your reference frame, against which the others will be compared. You choose a box size and pick the region for Registax to align the frames. Then press 'Align,' and it will find the corresponding region in all the other frames. Then you fiddle with the quality settings until you choose one quarter to one half of the frames, depending how good the seeing is that night. Now press 'Limit,' and only the best frames will be carried to the next step (Image by the author)

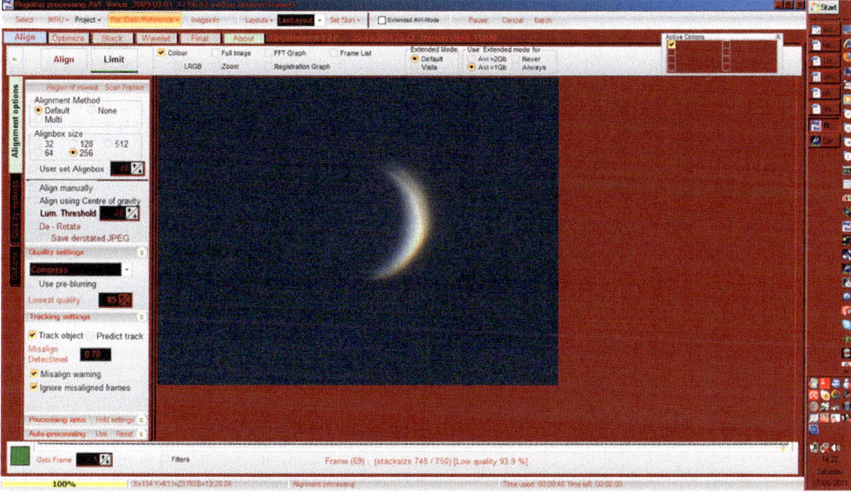

Fig. 4.10 Here is an example of a bad frame that Registax will reject (Image by the author)

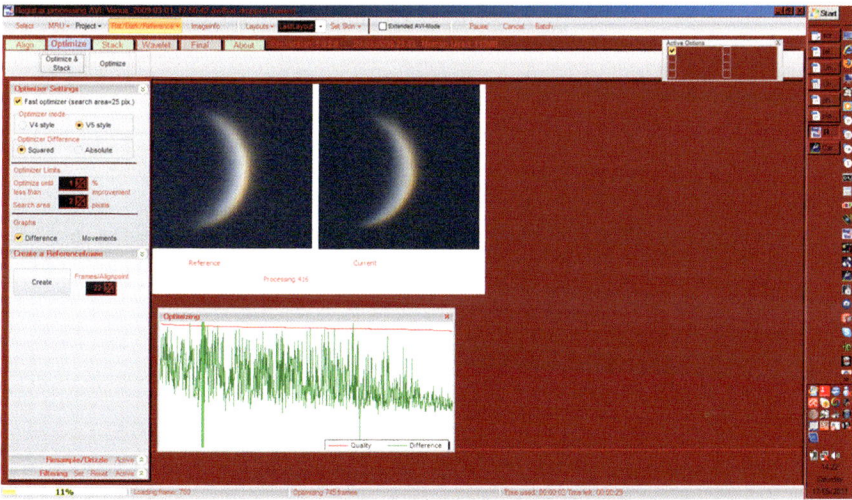

Fig. 4.11 Now press the 'Optimize and Stack' button, and Registax will fine-tune the alignment of the chosen frames. This can take a while. It then stacks the frames on top of one another. There are buttons you can fiddle with to get the best result here, but the default settings are usually quite good. What you will occasionally need to do is to go back if the red line on the graph shows that the quality of the images falls off a lot, and re-select fewer frames (Image by the author)

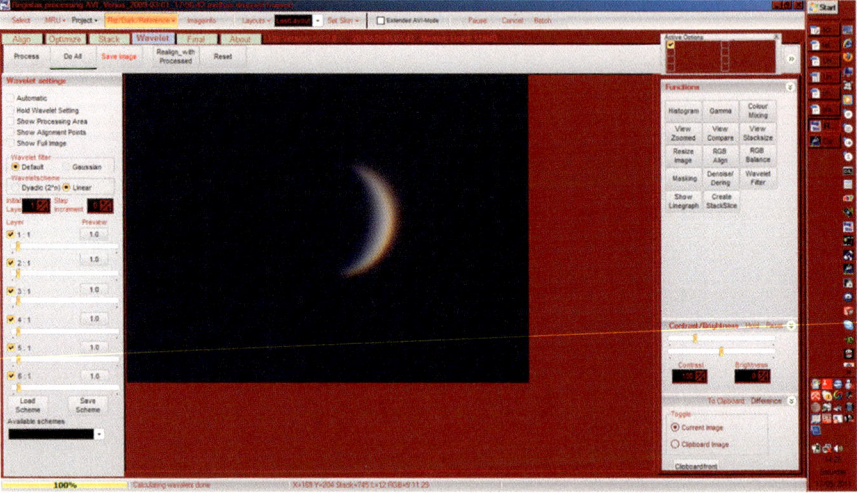

Fig. 4.12 The good frames have now been stacked. You are ready to bring out the detail in the image. There are red fringes to the right of Venus in this image, and blue fringes to the left. This is because of chromatic aberration in Earth's atmosphere. The effect is greater the closer you are to the horizon (Image by the author)

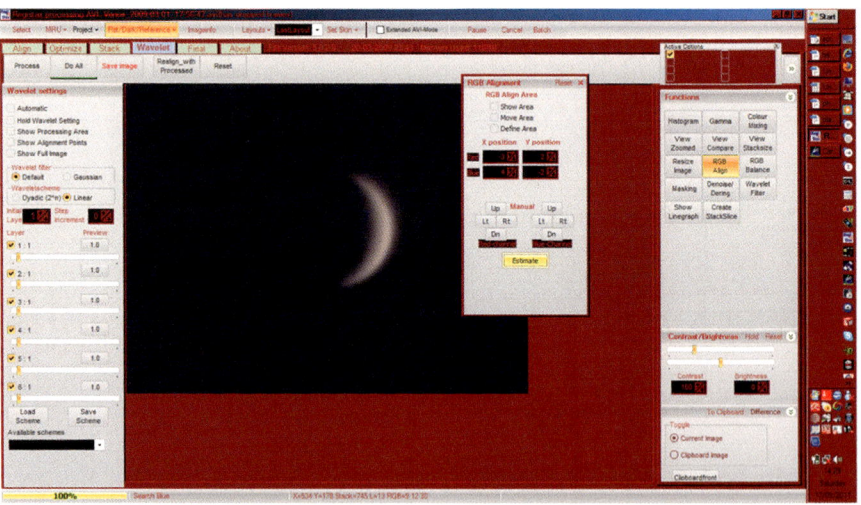

Fig. 4.13 Clicking on the RGB align button pulls up a menu in which you can correct for this. Generally, its automatic alignment ('Estimate') works very well, although you will sometimes need to override it. The image has now lost the color fringing (Image by the author)

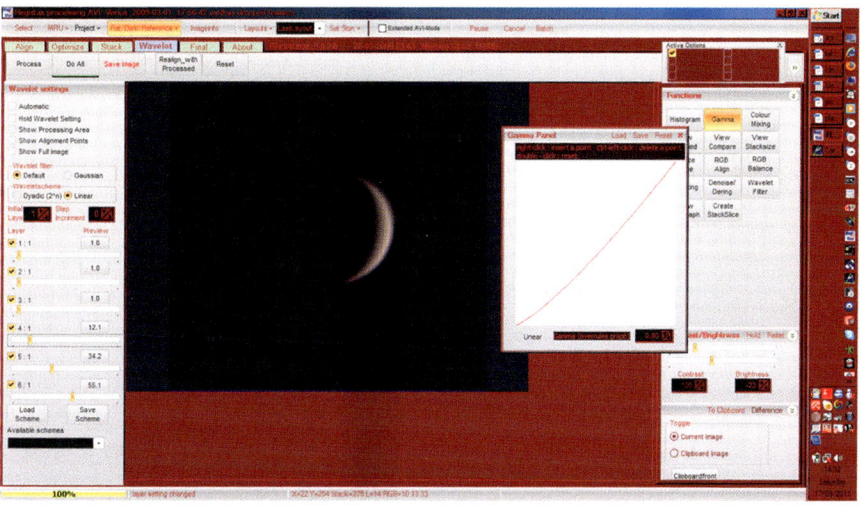

Fig. 4.14 Now you sharpen up the image. The only thing to do is to play with the wavelets until you learn what works. The settings shown work well for Venus, and for Jupiter's moons. Earth's Moon needs only gentle sharpening, but the atmospheric bands on Jupiter and Saturn need very aggressive sharpening. If you VERY slightly overexpose the avi movies, you can then reduce the value of gamma quite a lot, which enables you to sharpen the images more. You can also play around with the brightness and contrast. Once you are happy with the image, save it as a png file, which is the default (Image by the author)

Focusing

The plan for this chapter contained the words, "Include some remarks on focusing." It is going to be a lot harder to include some printable words about focusing. The "cussing" syllables in this word are the operative ones.

There's no doubt about it – focusing is a tough job. However, as with all awkward jobs, you can make life easier for yourself, or you can make life hard.

For visual observing it is not quite so bad because your eye will focus a slightly out-of-focus image – if you are young enough still to be able to focus, that is. At 50 you will have a lot less focusing range in your eyes than at 20. At 70, you will probably have virtually none.

The way to make life hard is to try to focus onto a camera chip by eyeball. One of two things will happen: you will get out-of-focus pictures or you will drive yourself crazy trying to focus. Or both.

As you become practiced you will become more and more aware of a slight misfocus. This may be the point at which to invest in, or make, a focusing aid. By then you will get the benefit because you will understand why the focusing aid helps.

Hartmann Mask

This is a rather grand name for a mask with three holes in it, which you hang onto the opening of your telescope. Figure 4.15 shows what a Hartmann mask looks like.

Figure 4.16a–c shows how the out-of-focus 'blob' on your CCD chip is split into three parts that look like three images. If you improve the focus, the three images will converge. If you worsen focus, the three images will move apart.

In practice, of course, you move them into and out of focus, hopefully getting better focus every time you do this. You train yourself to anticipate when the three images will cross, and to stop focusing at that point.

It is possible that the three images may not converge. This means that your telescope needs to be re-collimated. Indeed a Hartmann mask is a good way to test your collimation.

You need a very bright star to focus with a Hartmann mask. It does not work well with planets or the Moon. Find the brightest star you can. However, do make sure that the star is not a visual binary, such as Castor or Pollux; you will drive yourself crazy trying to make head or tail of six images that do not quite converge.

You can spend a fair bit of money buying a plastic Hartmann mask that you could easily step on and break in the dark, because they never seem to learn not to make these things entirely out of black plastic and are equally slow to learn that they should make them out of rubber so they survive being stood on. One answer to this is to make yourself a Hartmann mask out of cardboard and duct tape. It is not difficult.

You will need to set your CCD chip to its maximum brightness to use any focusing mask.

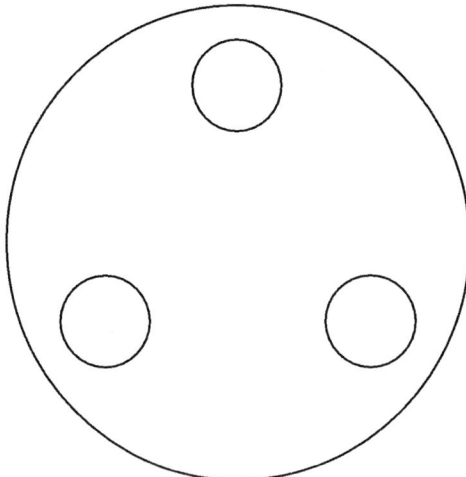

Fig. 4.15 A Hartmann mask is a mask with three holes in it. The holes need to be smallish, but not too small, relative to the telescope aperture. If your telescope is a reflector or a catadioptric, your holes need to miss the secondary mirror (Image by the author)

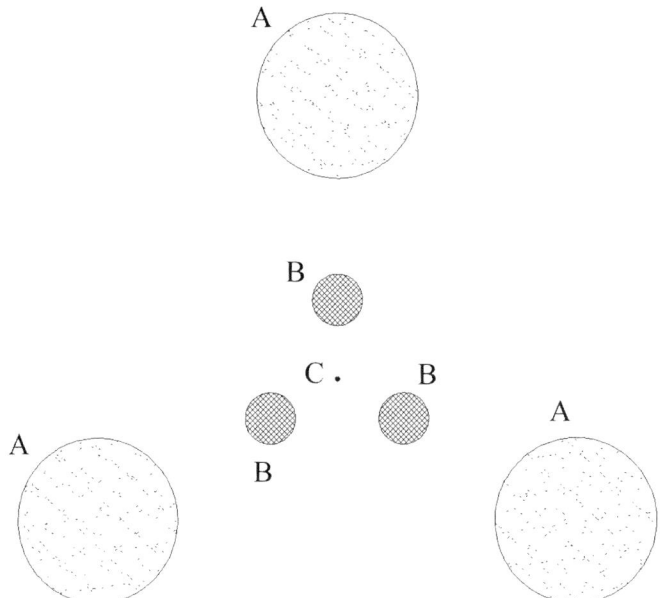

Fig. 4.16 (**a–c**) Images of a star when a Hartmann mask is attached. (**a**) Very out of focus. You see three fuzzy images that are far apart. (**b**) Slightly out of focus. The three images are getting sharper, and moving together. (**c**) In focus. The three images are sharp and on top of one another (Image by the author)

Bahtinov Mask

You pronounce this BACH-tin-of, the Bach pronounced like the composer.

This focusing mask is every bit as amazing as it is unpronounceable for an English speaker. It can be bought very cheaply on eBay. These masks are still black, but they do come with a reflective patch stuck to one side.

Bahtinov masks (Fig. 4.17) have to be made specifically for a given telescope type. You can buy them, but you can also download stencils from the Internet.

Bahtinov masks work by making three diffraction patterns. When the image is in focus, they cross. When it isn't, they don't. Figure 4.18, below, shows what happens to an image as the telescope is moved into and out of focus.

You do not need such a bright star as for a Hartmann mask. Jupiter's moons are about 5th magnitude, but you can focus them quite well with a Bahtinov mask and webcam – much better and more quickly than with the *Mark 1 Eyeball*. Figure 4.18

Fig. 4.17 A Bahtinov mask (Image by the author)

shows how this works. In the top two pictures, the image is being moved towards focus. The moons look a complete mess, and you can see faint, colored non-crossing lines around Jupiter. The bottom image is in focus. Now the moons are sharp points surrounded by diffraction lines. Jupiter is also surrounded by colored diffraction lines, which now cross one another. You can either align the diffraction spikes on Jupiter so that they cross, or you could focus until the moons look their sharpest. Once that happens, all the other stuff falls into place. The diffraction lines from the Bahtinov mask appear instead of the very confusing shapes where the moons should be in the out-of-focus pictures.

If you use the Bahtinov mask to focus a planet directly, there needs to be enough space on the chip for you to see the diffraction lines as well as the planet.

One of the more irritating aspects of focusing, especially with Newtonian telescopes, is that you have to keep moving from a convenient position to see your computer screen to operate the focuser. Electric motor-driven focusers are a real boon: you can focus while you are next to your laptop. They also reduce to virtually nil the time the telescope needs to settle down after re-focusing.

In general, you need to allow 15–30 min for re-focusing if you change cameras or add a Barlow lens. You can buy rings to put onto your focuser to reduce the need for continual re-focusing. There is also nothing to stop you using nail polish or something to paint a line on the focuser to indicate that you are almost in focus.

There are times when being way out of focus helps. If you are trying to find an object on your camera CCD chip, if it is way out of focus, it projects a bigger image onto the chip, making it easier to find. Better yet, use an eyepiece with cross-hairs of the kind used for drift alignment. This will greatly reduce the time you spend trying to place images of objects onto your webcam chip.

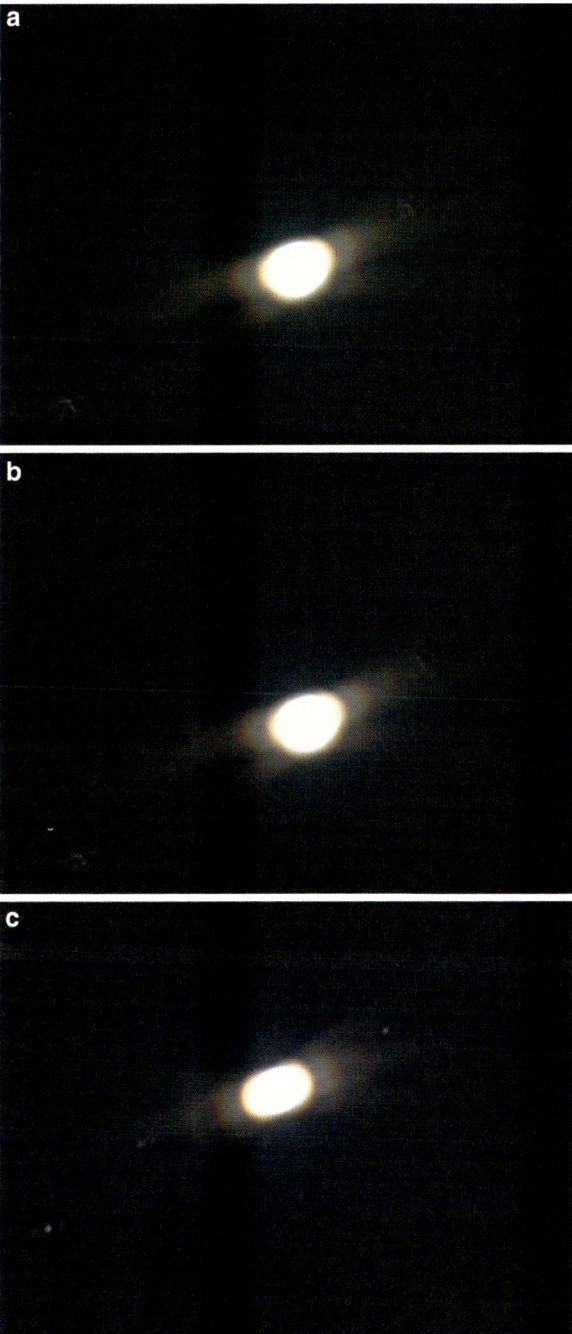

Fig. 4.18 Focusing Jupiter and its moons with a Bahtinov mask. There are three diffraction spikes, two which always make an 'X' and a third which we will call a 'cross-bar.' The cross-bar moves as you go into and out of focus (**a**) and (**b**). When it crosses the 'X' at its intersection point (**c**), your image is in focus (Image by the author)

Deep Sky Imagers

In general these are far more expensive than lunar and planetary imaging cameras, but they will work for Solar System imaging.

For example, you could have a monochrome Meade Deep Sky Imager to photograph Jupiter and its moons, with a deep green filter, because you can then bring out both the moons and the features on the planet, and the viewer is none the wiser about the color of your filter. This technique even works tolerably well in the summer, in places where there are very long dawns. As long as you can find the planet in the dark and keep track of it, you can follow it until sunrise. The moons are then swamped out by skylight, but you can still follow the planet. Venus is also relatively easy to follow in daylight. Again, the trick is to find it while it's still dark. A deep red filter, or even an IR filter if you feel brave, will get rid of a lot of skylight.

You could use a Meade DSI with *K3CCDTools* in much the same way as a webcam. You can also use it with its own software.

If you are using long exposures with a deep sky imager, it is particularly important to take dark frames and flat frames, or your image quality will suffer.

DLSR Astrophotography

Oh no, not another FLA! Another what? Four-letter acronym of course. The trouble is, digital single lens reflex camera is a bit of a mouthful. Why, of all the budget digital cameras you can buy, would you chose a vastly more expensive camera?

One of the main reasons is that you really want one where you can change the lens. With the relevant adaptor, your telescope could be the lens. Many people also piggyback their cameras onto their telescopes, to get the advantage of a mount that can follow the heavens. Figure 4.19 shows these configurations.

Fig. 4.19 The two main ways to mount a DSLR. (**a**) The telescope is the lens. (**b**) The DSLR uses its own lens and is piggybacked onto the telescope and mount. It is best not to make your dew hood out of white plastic; make it black instead. (**a**) The camera in the picture is actually an old 35 mm SLR. (**b**) This picture was taken with a telephone camera (Images by the author)

In principle you could put the DSLR straight onto the mount and forget the telescope. With a 200-mm focal length lens, this is a little awkward to do, because it's a pain to look through the viewfinder, but it works just fine.

However, if you want higher magnification and longer exposures than 10 s, you really need to fine-tune the motion of your mount. You do this by putting a second camera into your telescope, and using it and your laptop to guide the telescope.

What kind of camera do you need for this? If you are tracking a bright object like a naked-eye planet, you can use a webcam. If you want to track on a fainter object, you need either a deep sky camera or a guiding camera. These are generally lower quality than deep sky cameras, and hence are cheaper, but are capable of longer per-frame exposures than the $^1/_5$ of a second limit of a webcam.

You will need a second telescope, attached to your main one, to hold the tracking camera, and one to hold the main camera. Although it goes without saying that the telescope used to take the picture probably ought to be of good quality, frankly you can use almost any old trash for the guiding. An indifferent telescope will do just fine. It probably helps if it is not too heavy; otherwise you will have to spend the money you saved on the rubbishy telescope to buy a mount that can take the payload. The cost of mounts goes up a lot faster than the cost of telescopes as quality increases.

(Because this is true, it is impossible not to wonder why telescope makers don't put more effort into reducing the weight of telescopes. Could it perhaps be because they like making money out of selling you a second mount when you discover that the one they sold you with the telescope isn't good enough?)

There are plenty of free software packages that will guide your camera, such as *Guidemaster* and *PHD*. Both packages are rather awkward to learn, because they assume you know more than you probably do about your mount, so a tutorial is given later in the chapter.

DSLR File Types

If you try to perform manipulations such as subtracting dark frames on the JPEG files produced by DSLR cameras, you are in for disappointment. They do not work terribly well.

The trouble is, JPEG files are actually a kind of 'executive summary' of what you photographed. They don't contain all the information. They contain more than enough information for holiday snaps, but there is a 'native' file format for your DSLR camera. Collectively, these are called 'raw' files, sometimes RAW files, but there is no such thing as a file called 'image99.raw' or whatever. Raw is a generic name for native image file formats. For recent model Canon DSLRs, these files are called 'image99. CR2' or whatever; for Nikons they are called something like 'image99.NRF'. These two brands between them account for most DSLRs in astronomical use.

There's a lot more information in a raw file. You can't store nearly as many of them on your camera's memory chip as you can JPEG files. Left to itself, the camera will store a JPEG file and throw away the raw file. You can program it not to. Your manual will tell you how.

Raw files are proprietary. Although they have been somewhat reverse engi-
neered, so that they can be analyzed for astrophotography, you can't look up the file
details on the web; and the camera manufacturers keep moving the goalposts.
Attempts to standardize raw files have not been successful. Additional attempts are
probably not likely to be made. Why would Canon and Nikon want to teach their
less successful competitors such as Pentax how to catch up with them?

As soon as you start playing with raw files you will discover something interesting: a lot of the do-it-yourself manipulations have already been done when the
camera produces a JPEG file. For example, you can find a few hot pixels in the
JPEG files if you use *Registax* to sharpen the image, but you will find many more
in raw dark frames. The camera is already programmed to deal with them. The
colors in the JPEG images are also more realistic. This little machine processes
each shot to produce a JPEG file in a second or so. It contains a remarkable amount
of on-board intelligence.

Deep Sky Stacker

This free software package, written by Luc Coiffier, is very simple to use. It can be
downloaded from http://deepskystacker.free.fr. As its name implies, it is not written
for planetary imaging, so it does not attempt to rival *Registax* (Figs. 4.20 and 4.21).

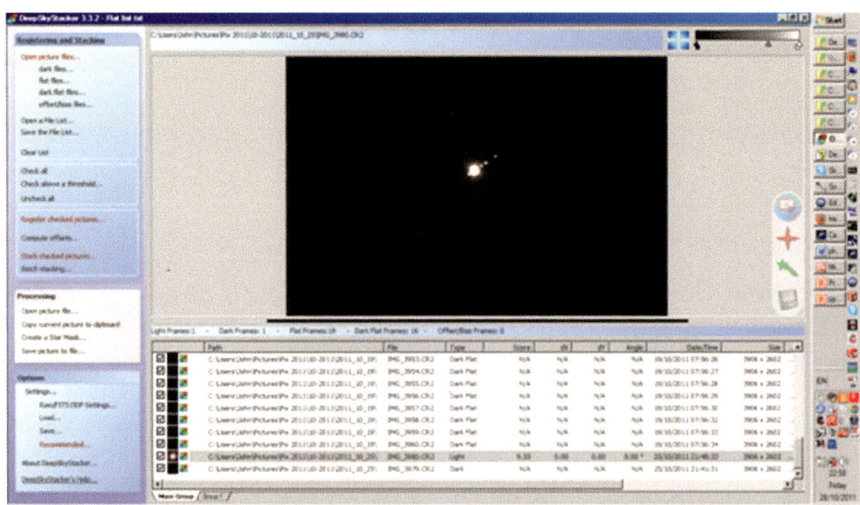

Fig. 4.20 The user interface for Deep Sky Stacker registering and stacking module (Image by
the author)

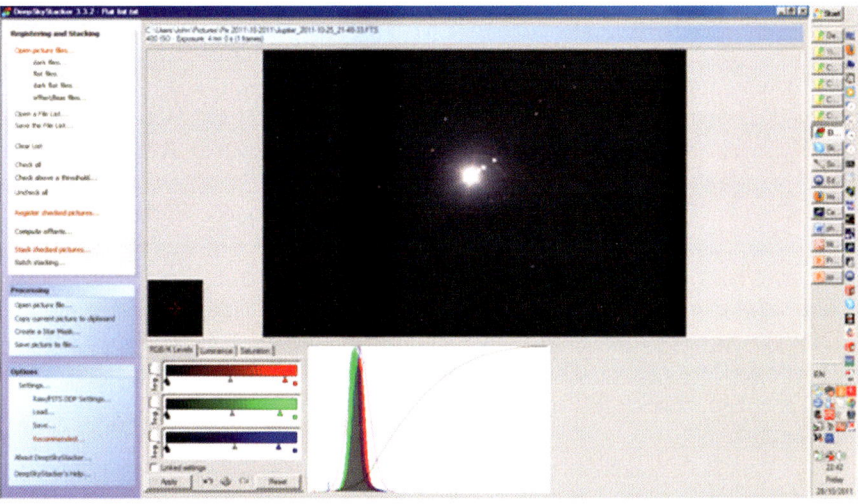

Fig. 4.21 The user interface for Deep Sky Stacker processing module (Image by the author)

Flat Fields

There are two reasons for taking flat field images. The first is that it is a great way to get rid of the annoying blotches left by dust on your images. The second is that it will compensate for an effect called 'vignetting,' whereby the image brightness falls off towards the edge of your image.

The latter effect is not that bad with a 640×480 webcam and an 8-inch f/6 telescope, because the chip is so small. It is *very* noticeable with a DSLR and a 500-mm telephoto lens.

They don't tell you this in most books, but there is another way to deal with dust bunnies. If your planet image drifts around the chip during the exposure, because your mount is not as well aligned as it might be and you are using your mount's hand controller to correct for this and keep the image on the chip, you can nudge it away from the worst dust bunnies, and the stacking process will deal with the rest.

Specialist books on astrophotography will advise you to build a light box to put over your telescope to take flat frames. You don't have to do this, although you certainly can. Unlike your dark frames, your flat frames do not need to be taken at the same temperature as your image. You can take them beforehand or afterwards. You are advised not to re-focus your telescope or move the camera in any way, which can be a little limiting. A diffuser in front of the image helps, but there are even (tedious) techniques for doing without one. You could place tracing paper over the 'scope or telephoto lens hood, and hold it on with an elastic band. How low-tech is that? You can either shine a light down the telescope or shoot your flat frames against the sky when it is not dark. Set your exposure so that the chip is about half-saturated, i.e., where it would be in auto-exposure mode.

Suppose that you were take your flat frames against the sky one morning, and the DSLR camera chooses an exposure time of 1/80 of a second. It tells you this in the viewfinder if you look up in the manual how to decode the flurry of numbers it throws at you. Don't forget that the manual is probably online if you have lost it. This information is also stored with both raw and JPEG files by most DSLRs. Most halfway decent image browsers will tell you the exposure time if you poke around a little to find it – look for something like 'properties' or 'info' in the menus. You need to know this because each of the 16 or so flat frames will need a dark frame, and with the lens cap on, the automatic exposure won't be able to choose an exposure time. You must set this manually; and it must be the same exposure time as the flat frame.

You will get a much better flat frame if you also shoot a dark frame for it. Do this at the same time as you take the flat frames so that the chip temperature is the same, and subtract the dark frame from the flat field image in the usual way. Packages like *Registax* or *Deep Sky Stacker* will do all this processing for you.

You need at least 16 flat frames and 16 dark frames to subtract from them. With a DSLR you need to take these manually. With a webcam, you need to produce some avi footage. A law of diminishing returns applies, but it is no big deal to shoot 150 flat fames and 150 dark frames for them. That's only 30 s exposure, each at 5 frames per second.

The benefit of all this hard work to generate flat frames is that your images will look a lot cleaner, and with the DSLR you won't get such bad vignetting. This is true despite the way the camera produces good JPEG files. You can do better if you are patient enough.

FITS Files

FITS files (.FTS) are the standard format used by professional astronomers. Consequently there is a good deal of software around that is capable of processing FITS files. The FITS file protocol is public information. You can manually include quite a lot of information about the picture. Figure 4.22 shows what the header of a FITS file looks like. You can add your own comments to the header.

The downside of using a professional format is that you can't read it with common amateur photo editors like *Photoshop Elements*. Not even *Registax* can read them. *Deep Sky Stacker* allows you to copy the processed image to the PC's clipboard. You can then paste this into the photo editor that comes with *Microsoft Office*, and save it as a bitmap file. From there you can continue to process the image in *Photoshop Elements* if you so wish. The file sizes get rather large, though. One image from *Deep Sky Stacker*, which started life as DSLR pictures, is about 40 MB, and the corresponding FITS file is about 60 MB.

```
SIMPLE   =                        T / file does conform to FITS standard
BITPIX   =                       16 / number of bits per data pixel
NAXIS    =                        3 / number of data axes
NAXIS1   =                     3906 / length of data axis 1
NAXIS2   =                     2602 / length of data axis 2
NAXIS3   =                        3 / length of data axis 3
EXTEND   =                        T / FITS dataset may contain extensions
COMMENT    FITS (Flexible Image Transport System) format is defined in
'Astronomy
COMMENT        and  Astrophysics',  volume  376,  page  359;  bibcode:
2001A&A...376..359H
BZERO    =                    32768 / offset data range to that of
unsigned short
BSCALE   =                        1 / default scaling factor
ISOSPEED=                        400
EXPTIME =                      240. / Exposure time (in seconds)
EXPOSURE=                      240. / Exposure time (in seconds)
SOFTWARE= 'DeepSkyStacker 3.3.2'
END
ƒ|‚ē,1‚ ƒ⊐‚ê,Ã,æƒ ‚ã‚¿ƒ4ƒ°‚#€œ⊐B⊐é,†ƒ&,œ,⊐,»ƒe,s⊐⊏,Rƒ$ƒû„Ó„Ô„Õ„¼„£ƒ¯…
hƒÅ,"ƒe„¨ƒ°‚‚ƒÐ„é,t€,d⊐Ê⊐1ƒ[…†…6„ç„⊐„⊐,\€œ,⊐ƒ„ƒDƒ⊐,²,`ƒ²…⊐ƒl⊐Ó⊐⊐€S,T„
Vƒ9

G„A„œ„ø…\…
```

Fig. 4.22 The first part of a typical FITS file, in this case written by Deep Sky Stacker (Image by the author)

The gobbledygook at the bottom is the binary information, i.e., the image. Of course in practice by far the bulk of the file is binary information. The information like exposure time was not manually supplied. This was downloaded from the camera as part of the raw image file it created:

Guiding: A Brief Guide for the Perplexed

The basic idea of guiding is that you point a camera at your photographic target, or a bright star close to it, and pipe the image live to your computer. Guiding software recognizes the target star, and perhaps even selects it, and sends instructions via a cable to your telescope mount to keep the guide star at the same place in the image. You usually have the guide camera look through a secondary telescope, which is mounted on, or next to, your main telescope.

Elsewhere, your main camera is looking through your main telescope, taking a photo, hopefully without blurring the image because your mount is tracking imperfectly – which they always do unless you spend an exorbitant amount on your mount. That's why you need to guide – to correct the tracking imperfections before they ruin your photos.

Unfortunately, this is a subject that is riddled with confusion. The following account is skeptical about a sacred cow or two. This is not a tutorial on deep sky photography – our subject is, after all, the Solar System. The methods suggested here work for Solar System photography. They can be used to measure the planets' positions to within an arc second or two of the published values, which is good evidence that they work well.

If you are taking webcam pictures at 5 frames per second, you do not need to guide. *Registax's* stacking will take care of the tracking imperfections. It may help you to guide if you are tracking the movements of the satellites of Jupiter or Saturn, or the rotation of Mars or Jupiter, by taking sequences of photos. Provided you are not expecting intermittent clouds to obscure the guiding image, and thereby cause the guiding software to lose track of its target, you could set such photography up automatically. Otherwise, for webcam photography, don't bother to guide.

There are several free guiding software packages, such as *PHD* or *Guidemaster*. *PHD* allegedly stands for 'push here dummy,' implying that you do not need a real Ph.D. to operate it. You don't, of course, but in fact none these packages is terribly easy to learn. They are not as intuitive as some people would have you believe.

In particular, the questions they ask assume quite a high level of background knowledge about the electronics, software, and firmware in your mount. The first thing you need to work out was what telescope type the guiding software thinks your mount is. For example an HEQ5 with the so-called Rajiva update, an electronics control system you can buy quite cheaply, emulates a Meade LX200 mount, as it does many other mounts. You really need to be prepared for a certain amount of trial and error to find out which protocol your mount uses if the manual does not tell you.

You also have to feed the software a little information about your telescope – the focal length of the objective, for example. *Guidemaster* also asks you a whole lot of other questions about your telescope you haven't an earthly chance of answering. Ignoring them and hoping and using the defaults will probably work.

It is worth rotating your guide camera until the right ascension control moves the image in the guide camera horizontally and the declination control moves it vertically. The guiding will then work much better; otherwise it will not understand which movement is right ascension and which is declination. The correct orientation depends how you rotate the telescope tube in the mount for a Newtonian. It can be surprising what this orientation actually is.

You don't need brilliant focus to guide with a bright object such as a naked-eye planet. Nor do you need great collimation. You will find guiding easier if you do focus and collimate properly. In general, people guide using stars, which are quite faint. Then you really do need very good focus and collimation, or the software won't recognize the star, especially if a bit of thin cloud drifts in front of it.

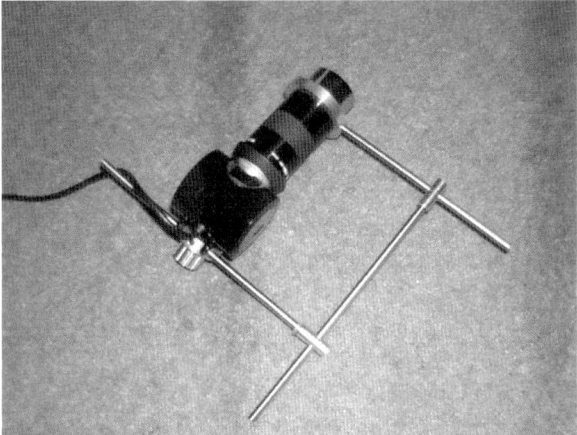

Fig. 4.23 A webcam in afocal projection. It sits behind a telescope eyepiece just as a human eye would. The arrangement of rods and grub screws is used to hold the camera onto the eyepiece (Image by the author)

For guiding on naked-eye planets, the author uses *Guidemaster* with a webcam in afocal projection, i.e., keeping the built-in lens on the webcam and holding it onto an eyepiece with a mechanical fastening (Fig. 4.23).

Webcams are not powerful enough to use stars for guiding. The problem is that you need a longer exposure than $1/5$ of a second. You can still buy second-hand modified webcams for guiding. There is a certain amount of anecdotal evidence that they can be prone to 'hanging' and ruining photographs because they 'lose the plot.' A dedicated guiding camera will cost at least twice as much as a used modified webcam. New 'modded' webcams are no longer easily available.

One subject mentioned in books and on websites in reference to guiding is periodic error correction. Ultimately it is all about correction for the fact that your mount is driven by gear wheels, which inevitably have geometric and mechanical imperfections. These imperfections manifest themselves once per gear wheel revolution. Unless you are lucky enough to have a very high quality mount permanently installed on a pier in a concrete foundation, there is no point in bothering with this for Solar System imaging. If the guiding works properly, it will simply deal with this phenomenon, along with the imperfections of tripod mount misalignment. In any event the cheapest mounts with guiding capability are not necessarily going to be good enough to have reproducible periodic errors.

Referring to Fig. 4.24, in your guide camera image, you pick a bright star, click the 'guide' button, and *Guidemaster* offers you three choices. It will adjust until the guide star (or planet) is centered, keep it in its current position, or move it to the last position to which you guided it.

You can put the cap on the guide scope and shoot a dark frame if you wish. For guiding on every planet except perhaps Venus and Jupiter, it is well worth while shooting a dark frame.

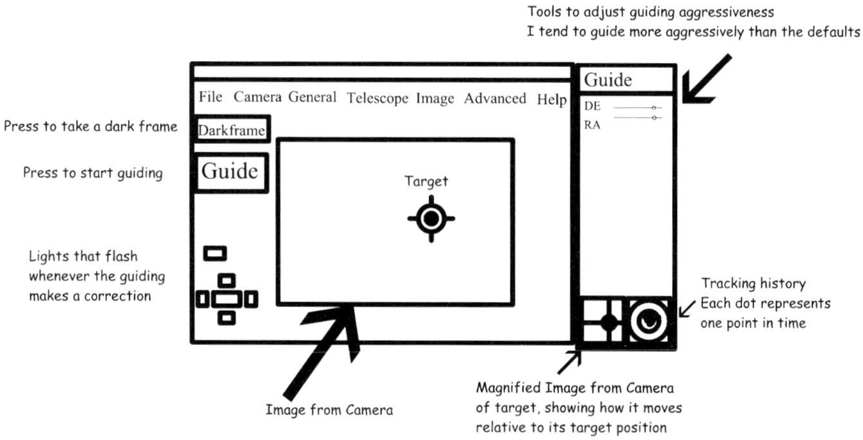

Tools to adjust guiding aggressiveness
I tend to guide more aggressively than the defaults

File Camera General Telescope Image Advanced Help

Guide
DE
RA

Press to take a dark frame
Darkframe

Press to start guiding
Guide

Target

Lights that flash
whenever the guiding
makes a correction

Tracking history
Each dot represents
one point in time

Image from Camera

Magnified Image from Camera
of target, showing how it moves
relative to its target position

Fig. 4.24 Sketch of the main guiding features of Guidemaster (Image by the author)

How aggressively should you guide? The resultant pictures are noticeably sharper if you guide very aggressively.

The 'camera' menu is for you to select which camera to use for guiding. It offers you whatever cameras are on the computer, including the built-in webcam if there is one.

You are likely to discover that your mount has a 'sweet zone,' not too far from due south (north in the Southern Hemisphere), not too high and not too low. In this sweet zone, the guiding will give you pinpoint stars. Outside it, your stars will look more like lines. The size of the sweet zone depends in part on whether your mount is overloaded with equipment.

Chapter 5

The Solar System
in Context

The Big Bang

The known universe is expanding. Our evidence for this comes from the fact that astronomers can measure the distances to other galaxies, and can measure whether they are approaching us or receding from us, and if so, how fast. What is found is that nearby galaxies may be approaching one another, but that on scales bigger than that of galactic clusters, the galaxies are receding from one another. The further away they are, the more they recede.

Where does this evidence come from? Evidence for distance is gathered by finding 'standard candles,' objects whose absolute brightness is known from other measurements, and by assuming that their perceived brightness falls off as the square of their distance from us. That is what the brightness of a car taillight does as the car moves away. Why the square? Because the apparent brightness is proportional to the area over which the light spreads, as shown in Fig. 5.1.

The speed at which the galaxies recede is determined by the Doppler effect (Fig. 5.2). This is the effect according to which an approaching vehicle's noise is higher pitched than that of the same vehicle when it recedes. The wavelength of low-pitched sound is longer than that of sound at a higher pitch. The same is true for light and other electromagnetic radiation, as shown in Fig. 3.1. Lower wavelengths of light mean redder light. Hence the light from a red object is said to be red-shifted.

The further away a cluster of galaxies is, the more its light is red-shifted. This effect was noted by Hubble in the 1920s [43]. This effect is illustrated in Fig. 5.3.

© Springer Science+Business Media New York 2015
J. Clark, *Viewing and Imaging the Solar System*, The Patrick Moore
Practical Astronomy Series, DOI 10.1007/978-1-4614-5179-2_5

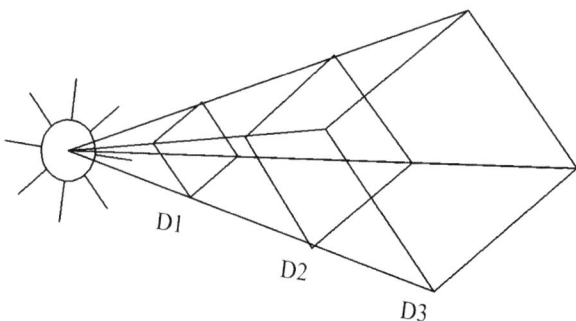

Fig. 5.1 The same amount of light from the celestial object has to be spread over each of the areas at distances *D1*, *D2* and *D3*. These areas are proportional to *D1* squared, *D2* squared and *D3* squared. Thus the apparent brightness of the celestial object falls off proportionally to the square of the distance (Diagram by the author)

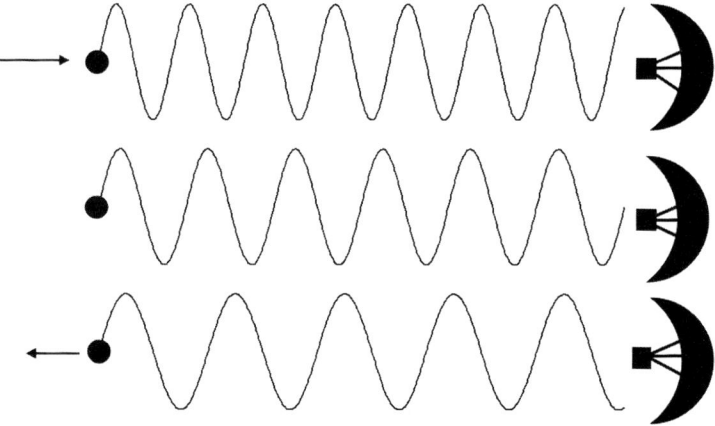

Fig. 5.2 *Top*: If a wave source is moving towards the detector, the wave appears to be compressed. It has a higher frequency and shorter wavelength than if the source were stationary relative to the detector (*center*). If the source is moving away from the detector (*bottom*) the wave appears to be stretched. It has a greater wavelength and lower frequency than in the static case (*center*). This is called the Doppler Effect (Image by the author)

Measurements of the velocity of recession, known as the Hubble constant, have been refined ever since.

Figure 5.3 begs the question – when were all the galaxies in the same place? The answer appears to be about 13.8 billion years ago. This time is known as the Big Bang and is taken to be the origin of the universe as we know it.

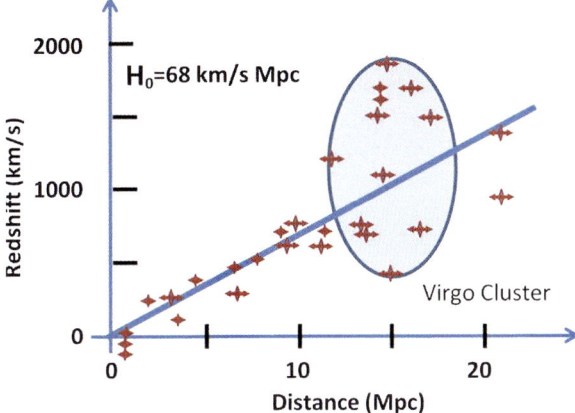

Fig. 5.3 Red-shift of galaxies versus distance. One Mpc is one megaparsec, a distance of 3,261,560 light years. This is a little over half as far again as the distance to the Andromeda Galaxy. The universe is expanding at a rate of 68 km per second for every megaparsec (Image courtesy of Brews O'Hare, http://en.wikipedia.org/wiki/File:Hubble_constant.JPG. Licensed under the Creative Commons Attribution-Share Alike 3.0 license)

The Evolution of the Universe

The question of what happened 'before' the Big Bang is complicated by the fact that this is the point at which time began. It appears that there was no 'before' [44].

If we make two assumptions, we can deduce quite a bit about what happened immediately after the Big Bang. The first assumption is homogeneity – that the universe is the same everywhere and in all directions. The second is that the laws of physics as worked out here on Earth apply everywhere. Since we cannot travel to distant galaxies to perform experiments to test these assumptions, we are forced to guess that they hold good. We can have a good look to see if we can disprove them, and we have, but that does not prove them. The cosmic microwave background radiation is the same in all directions to at least one part in 100,000 [45]. The physical force we can detect over the largest distance is gravity. Our best theory of gravity, Einstein's theory of general relativity, has so far passed every test we have tried [46].

If we assume that our nuclear physics has always been the same everywhere, we can use thermodynamics to deduce the abundance of the very lightest chemical elements, hydrogen and helium. In a popular book with minimal mathematics that became a bestseller [47], Steven Weinberg works through the argument, and shows how the cosmic background radiation got its temperature (around 3 °C above absolute zero) and how the ratio of hydrogen to helium is about 74 % hydrogen and 26 % helium, which is almost exactly what we observe.

Galaxies

Because light can only travel at about 186,282 miles per second, any information from galaxies, say 13.2 billion light years away, left them 13.2 billion years ago, i.e., a 'mere' 500–600 million years after the Big Bang. The most distant observed galaxy to date is in fact at about that distance from us, and formed 480 million years after the Big Bang [48]. We therefore know that galaxies have existed for most of the life of the universe. It is generally assumed that they separated from one another because of tiny fluctuations in the 'primordial soup' of the Big Bang. Shortly after this, the rate of star formation increased hugely.

Stars

Stars are thus a very old feature of the universe. They are balls of hot material, mostly hydrogen and helium (excepting white dwarves and neutron stars), kept roughly spherical by their own gravity. The gravitational pressure causes the initial heating, rather as the pumping up of a bicycle tire gradually heats up the pump. The electrons are stripped off the hydrogen atoms, which normally consist of one proton and one electron. As the pressure increases, some of these loose protons begin to undergo nuclear reactions, converting hydrogen to helium, and let out a tremendous amount of energy. Indeed, this nuclear reaction is the one that happens in hydrogen bombs, the most terrible weapons known to humankind.

The bigger the star, the faster it burns its hydrogen, because its self-gravity is stronger. The very biggest stars last only a few million years. Ones the size of our Sun last billions of years. Eventually, they run out of hydrogen. The story of stellar evolution is now well established [49]. Once their cores run out of hydrogen, the stars begin to form other chemical elements. When the helium begins to undergo nuclear reactions, it does so much more explosively than the hydrogen burning. This phase of the star's life is called the helium flash. The star expands, and helium turns to carbon and oxygen. Stars the size of the Sun will never get beyond converting helium to carbon and maybe oxygen. They will blow off their outer regions into red giants, leaving a core of carbon and possibly oxygen, which collapses into a white dwarf, a star with about the mass of the Sun squashed into about the volume of Earth. White dwarf material therefore has some very exotic properties.

More massive stars will keep generating the first 28 chemical elements in the Periodic Table, i.e., those up to iron. Nuclear reactions to make the other 64 chemical elements found on Earth absorb more energy than they give out, so they cool the star. In most such stars the nuclear reactions reverse and the star dies quietly. They may even become black holes, i.e., objects with an unattainable escape velocity faster than that of light. There is a mass range at which the cores of the stars become neutron stars. A neutron star has fairly precisely 1.44 times the mass of the Sun. It is so dense that it is only about a mile across, and has even more exotic properties

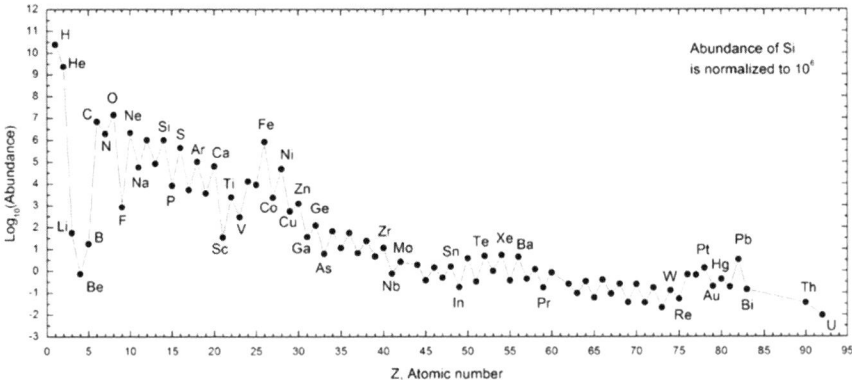

Fig. 5.4 Relative abundance of chemical elements. The atomic numbers along the horizontal axis are those found in any Periodic Table of the chemical elements. The symbols are standard chemical symbols, H=hydrogen, He=helium, C=carbon, O=oxygen, Fe=iron, U=uranium, etc. The vertical axis is the logarithm of abundance. The logarithm of the abundance of silicon is arbitrarily set to +6. A log abundance of +7 is 10 times this, +8 is 100 times more abundant, +9 is 1,000 times more abundant, etc. Similarly a log abundance of +5 is only 1/10 the abundance of silicon, +4 is 1/100 the abundance, etc. Thus if you do the math, you find out that hydrogen is over a trillion times as abundant as uranium (Image courtesy of http://en.wikipedia.org/wiki/File:SolarSystemAbundances.png, licensed under the Creative Commons Attribution-Share Alike 3.0 Unported)

than a white dwarf. The outer material cannot collapse into the ultra-dense neutron star, and it appears that the shock wave of the collapse rebounds, and causes the outer material to be expelled into a large cloud. The Crab Nebula, Messier object M1, is exactly such a supernova remnant with a neutron star at its core.

In hydrogen bomb tests, tiny traces of many chemical elements more complex than iron have been found. It may safely be assumed that these chemical elements are found in supernova remnants. Indeed, that is where most of the chemical elements other than hydrogen and helium come from. Nebulae such as M1 collapse under their own gravity and are recycled as stars (Fig. 5.4).

The story of how the chemical elements are formed in supernova explosions was worked out by Margaret Burbidge, Geoffrey Burbidge, William Fowler and Fred Hoyle in a landmark paper [50], which is so well known that it is usually referred to as the B²FH paper.

Planetary Systems

Anything written here about planetary systems seriously risks being out of date by the time of publication, so fast is this field of knowledge evolving.

Suffice it to say that the first confirmed exoplanet (i.e., a planet outside the Solar System) discovery was in 1988. At the time of this writing (February 2014) the number of known exoplanets is just shy of 2,000.

The easiest exoplanets to detect are those with very short orbital periods and those with high mass to make their stars 'wobble' as they orbit. Thus we would not have detected a planet like Saturn, which is about 10 AU from the Sun and has an orbital period of 29 years. There is considerable 'sampling bias' in what we have detected.

We still do not know whether most stars have orbiting planets, and whether the Solar System is especially unusual.

Stay tuned.

Chapter 6

Earth's Nearest Neighbor: The Moon

You can actually tell that the Moon is our nearest neighbor – at least occasionally – because as it makes its orbit, it passes in front of the planets and stars, etc. You never see anything pass in front of it. These phenomena are called occultations. They look quite good through a modest telescope. You don't see much with the naked eye. The astronomy magazines usually forecast occultations. An example is shown in Figs. 6.12 and 6.13.

The Moon is a great target to start off with, because it is trivial to find, easy to observe and by far the easiest target to photograph.

The Moon makes a complete orbit of Earth about once a month. You can watch this happen. If you get a clear spell of weather, note whereabouts in the sky the Moon is. Note also the time. The next night, note where it is at the same time. It will have moved east. New Moons are always visible near the glow of dusk. Full Moons are highest in the sky at midnight (midnight proper local time, not the irritating daylight savings or summertime that gets wished on us.) The waning Moon is visible in the wee hours. As you become more familiar with the constellations, you can watch how it passes through them over a month. You don't need massively detailed knowledge of the sky – just a reasonable eye for the brightest stars so that you have a vague idea which stars are where.

The Moon in fact moves east relative to the starry background by about one lunar diameter per hour. Over an evening's observing, it moves very noticeably to the east. Compare Figs. 6.12 and 6.13 below, where this effect is shown.

© Springer Science+Business Media New York 2015

J. Clark, *Viewing and Imaging the Solar System*, The Patrick Moore Practical Astronomy Series, DOI 10.1007/978-1-4614-5179-2_6

Fig. 6.1 On the *left*, the situation in the northern winter. The observer in the Northern Hemisphere sees the Sun low in the sky during the day, but the full Moon high in the sky at night. On the *right*, we see the situation in the northern summer. Now, the observer in the Northern Hemisphere sees the Sun high in the sky during the day, but the full Moon low in the sky at night. A similar phenomenon occurs for Southern Hemisphere observers. They also see a high full Moon in winter and a low full Moon in summer (Image by the author)

Another phenomenon you can observe without instruments is shown in Fig. 6.1. This figure assumes that the Moon goes around Earth in the same plane as the one along which Earth goes around the Sun. This is approximately true, but out by a little over 5°. Anyway, the diagram explains why you see full Moons very low in the sky in summer and high in the sky in winter. That's why bright moonlit nights are cold; they don't occur in summer except at the equator, where it is never cold anyway unless you are very high up.

Also, because the sunlight is weak in winter well away from the equator, the Moon is then often very visible to the naked eye. You don't see full Moons during the day for the obvious reason that the Moon then has to be behind Earth when looking from the Sun. But you do see other phases. This can in fact be a very good time for lunar observing with a telescope because the Moon is not then blindingly bright. It can be a bit blinding at night, so much so that people often use neutral density filters to make it dimmer. They make lunar viewing through anything more powerful than 7×50 binoculars much more comfortable.

People are often tempted to photograph the Moon with just their cameras. Those beautiful pictures of the Moon looking very large in movies rather encourage this. This is no bad thing to do, but a few things will improve your result no end. Those romantic shots in movies are taken with the kinds of telephoto lenses the paparazzi use surreptitiously to photograph the doings of celebrities. If you point your telephone camera at the Moon it will come out as a tiny dot. The same applies if you use a simple digital camera.

The second trick is not to overexpose the Moon. If you can control the exposure, do. Experiment with reducing the exposure until you see detail. Figure 6.2 shows an example. It was taken with a cheap 500-mm telephoto lens for a DSLR, which was obtained second hand. There are plenty of these available all the time, because the camera shops sell them to unsuspecting customers for photographing birds. If you read the Internet reviews, you quickly find out that too many people, spoiled by auto-focus lenses, get the shock of their lives when they find they

Fig. 6.2 The Moon photographed on May 6, 2009, 21:48 UT, using a DSLR with a cheap 500-mm catadioptric telephoto lens with a fixed f/8 aperture. You are advised to experiment with exposure times. This one was 1/2,000 of a sec. This picture is heavily cropped. Most of the image was *black* (Image by the author)

actually have to turn a knob to focus their cameras. So eBay and similar websites are swamped with these lenses going dirt cheap. They are not really difficult to focus on birds during the day.

Focusing for astrophotography is another matter. The focus in Fig. 6.2 is not great. It takes long time and a lot of patience to learn to focus this lens for astrophotography.

For Moon shots like this, by the way, there isn't much point. There are much better ways to obtain sharp pictures of the Moon.

In fact the telephoto lens is focused by exploiting its aberrations. You really need a DSLR with live view focusing, which older models lack, where you can focus using the image on the screen, and which you can pipe to your laptop if you want. You can magnify the center of the image by up to 10× for this purpose. If a planet such as Mars or Jupiter is around, it is bright enough to see in the 10× region, but preferably not quite at its center. Then, at the point of focus, the astigmatism in the image disappears. One side of focus the planet is oval and taller than it is wide. The other side of focus, it is oval and wider than it is tall. With practice you can find the cross-over point. When you first start, this may well take you half an hour, lots of trial images and even more cursing; but it will eventually only take you about 10 s.

Fig. 6.3 Moon, December 19, 2008, 05:02 UT. Canon EOS 100D DSLR, on an HEQ5 mount with no telescope, 200 mm f/5.6 lens, 1/320 s (Image by the author)

Fig. 6.4 Moon, December 19, 2008, 04:53 UT. Canon EOS 100D DSLR, on an HEQ5 mount with no telescope, 200 mm f/5.6 lens, 1/2 s. Towards the *top*, slightly to the *right*, a star is visible. It is Regulus (Image by the author)

Since the photo in Fig. 6.2 was exposed for only 1/2,000 of a second, there is absolutely no benefit to putting the camera on a mount and tracking the sky. You could probably even hand-hold the shot, although to suggest this is a little risqué.

Figure 6.2 shows some craters, the white blobs; and the rays coming out from the big crater at the bottom – a 50-mile-wide crater called Tycho – which give the impression that the Moon is a sphere. The dark patches called 'seas' are quite visible, and do not look like a face. So this rather basic picture does show more than you can see with the naked eye.

What can you see with a slightly less exotic telephoto lens? Figures 6.3, 6.4 and 6.5 show the kinds of results you can obtain. The Moon is quite small in the image

Fig. 6.5 Moon, December 19, 2008, 04:50 UT. Canon EOS 100D DSLR, on an HEQ5 mount with no telescope, 200 mm f/5.6 lens, 30 s. Several stars are now visible, but you can no longer see any detail of the Moon (Image by the author)

even at 200 mm focal length. These pictures were taken at various exposures to show you what would happen. A motorized mount was used this time. Quite a few of the nearby stars are in some of the images.

A few months later, when the Moon was again next to Regulus, the photo in Fig. 6.6 was taken. In both cases, two stars are visible, Regulus (α Leonis) and 31 Leonis. Since these stars are much, much further away than the Moon, we can safely say that when these two photos were taken, the direction from Earth to the Moon was the same in both cases.

Yet the phase of the Moon is rather different. The direction from the Sun to the Moon has changed, because both Earth and the Moon are orbiting the Sun. Also, the fact that the Moon is in the same position relative to Earth means that in the time between the photographs it had completed an exact number of orbits. In fact it had completed four orbits. If you do the sums, this turns out to be over a period of 109.77 days. The orbit time is therefore one quarter of this, or 27.44 days. That's about 0.4 % higher than the published value. So a very simple bit of observation gives us a really accurate value of the orbital period of the Moon.

Furthermore, you can also see something else in these pictures. The Moon has not changed its orientation relative to the stars in the 4 months between the pictures.

Figure 6.7 shows why the phase of the Moon is different after exactly four orbits of Earth. The phase is determined by how much of the side of the Moon currently pointing towards the Sun we can see, not by the direction of the line from Earth to the Moon. Thus the time to orbit is not quite the same as a 'lunar month,' the time taken for a given phase to reappear.

It is well known that the Moon presents pretty well the same face to us all the time because its rotation period is the same as its orbital period.

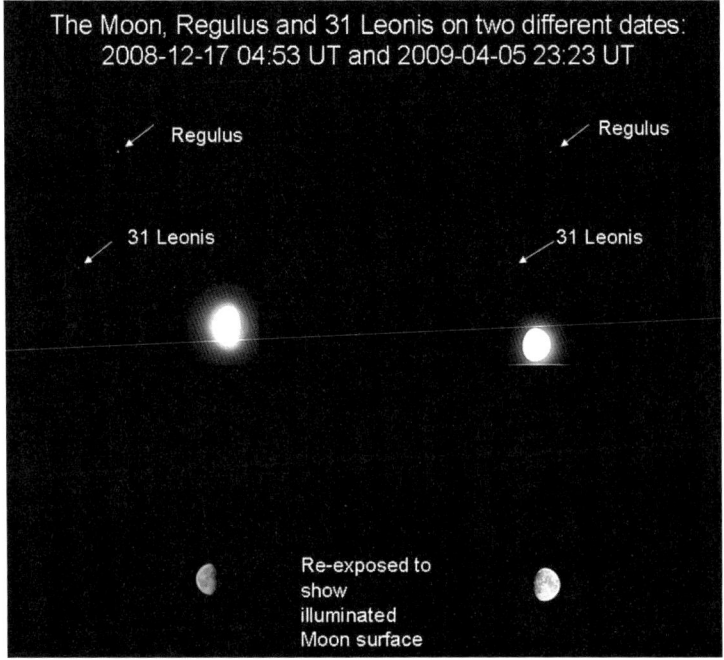

Fig. 6.6 Moon, April 5, 2009, 23:23 UT, plus the image from Fig. 6.4. Canon EOS 100D DSLR, on an HEQ5 mount with no telescope, 200 mm f/5.6 lens, 1/8 and ½ of a sec respectively. Two stars are visible, Regulus (α Leonis) and 31 Leonis. Since these two stars are much, much further away than the Moon, we can safely say that when these two photos were taken, the direction from Earth to the Moon was the same in both cases (Image by the author)

This observation contrasts with a simple naked-eye observation you can make over the course of the time from moonrise to moonset. Note the apparent orientation of the Moon. It changes as the night proceeds. In the Northern Hemisphere it appears to rotate clockwise. In the Southern Hemisphere it appears to rotate anti-clockwise. This simple observation proves that the 'field rotation' of the Moon is an illusion because Earth rotates. You might argue that this also has to be true because the Moon cannot simultaneously rotate clockwise for an American and counter-clockwise for an Australian. You would be right (Fig. 6.8).

It is well known that the Moon presents pretty well the same face to us all the time because its rotation period is the same as its orbital period. This is not quite true. Murray and Dermot [51] show out that the Moon in a tidally locked orbit actually points approximately at the unoccupied second focus of the Moon's elliptical orbit (Fig. 1.2), not the focus where Earth sits. We therefore see about 60 % of the Moon's surface over the course of an orbit.

What about the movement of the terminator not from one night to the next but over the course of an evening? Figures 6.9 and 6.10 show, upon close examination, that the progress of the terminator is slow but noticeable.

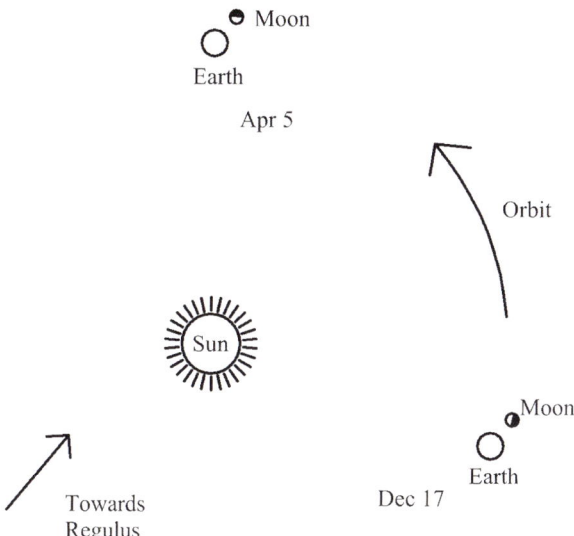

Fig. 6.7 Why the phase of the Moon is different after exactly four orbits of Earth. The phase is determined by how much of the side of the Moon currently pointing towards the Sun we can see, not by the direction of the line from Earth to the Moon. In the December 17 position, the illuminated part of the Moon is to the *left* and gibbous as viewed from the Northern Hemisphere – we are looking down onto the Sun's Northern Hemisphere. In the April 5 position the illuminated part of the Moon is to the *right* and gibbous as viewed from the Northern Hemisphere. The photographs were taken in England (Image by the author)

• The Orientation of the Moon can be measured

- Verify by checking lines joining points at edges of 'seas'.
- These lines are parallel.
- Photos in previous figure were aligned to make the line between Regulus and 31 Leonis parallel.

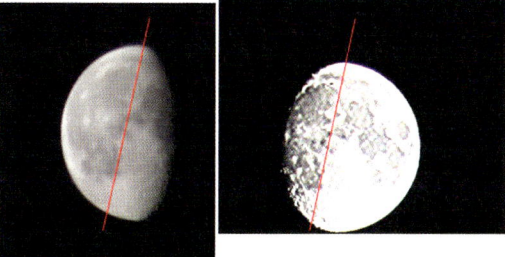

• The Moon has not changed its orientation relative to the stars.

Fig. 6.8 From the pictures in Fig. 6.6 you can check whether the orientation of the Moon has changed noticeably over 4 months. It has not (Image by the author)

Fig. 6.9 Progress of the Moon's terminator over 2¼ h. Close examination of these pictures shows slight but noticeable progress of the terminator (Image by the author)

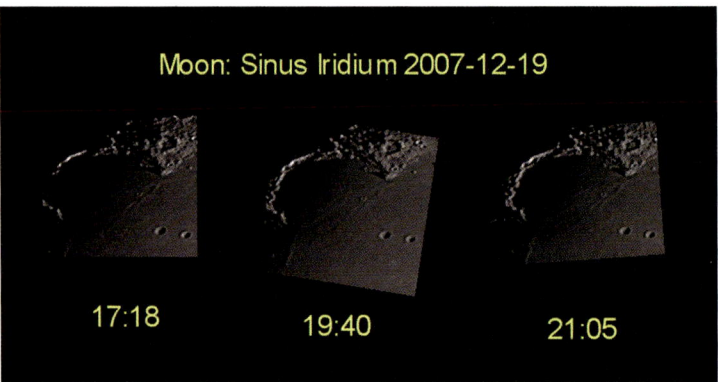

Fig. 6.10 Progress of the Moon's terminator over 3¾ h. Close examination of the tall mountain at the right hand end of the horseshoe ring of mountains shows that its shadow has moved. It can also be seen that the horseshoe ring of mountains has become more visible (Image by the author)

What we do not see is the side of the Moon shown in Fig. 6.11. You can only see this by looking from space. This side of the Moon was completely unknown until 1959, when a Soviet space probe flew past it and sent back pictures. That's why the one 'sea' on that side of the Moon was named the Sea of Moscow. It was a surprise that there are almost no seas on the far side of the Moon. Incidentally it is a mistake, albeit a popular one, to confuse the far side of the Moon with the dark side. The far side is the one that never points towards Earth. This side is light half the time, and

Fig. 6.11 The far side of the Moon. This photo is clearly a composite. The shadows in the craters are on different sides in different places (Image courtesy of NASA)

dark the other half. The Moon points the face we know and love towards Earth, not the Sun. It too is illuminated half the time, and half the time any point on the equator is part of the 'dark side.'

In an 8-in. telescope you can see the non-sunlit part of the Moon for about 4 days after a new Moon. This part of the Moon is lit by reflected light from Earth. You can see an example of this in Fig. 6.12, a shot of the Moon near the Pleiades, taken with just a camera.

Notice how the Moon has moved relative to the Pleiades in the hour or so between Figs. 6.12 and 6.13. Telescope mounts are normally set to track the stars, not the Moon. The more sophisticated ones can be set to track the Moon or even the Sun instead.

Binoculars and the Moon

So far we have barely mentioned telescopes or binoculars. The Moon is a superb observing target through binoculars. You can see the 'seas' and you can see craters with modest, hand-held binoculars. Also, very spectacularly, the Moon looks like a sphere through them. At the time of writing, there is a bit of a craze for lists of things to do 'before you die.' (Lists of things to do afterwards are rather shorter, but that's another story.) If there is one thing every sighted person should do before they die it is to look at the Moon through binoculars.

Fig. 6.12 The Moon close to the Pleiades, April 26, 2009, 20:34 UT. Canon EOS 1000D DSLR, 200 mm lens, 2.5 s at f/5.6. The main stars of the Pleiades have been inked in with *Photoshop* to compensate for the inevitable loss of quality in book production (Image by the author)

Fig. 6.13 The Moon close to the Pleiades, April 26, 2009, 21:39 UT. Canon EOS 1000D DSLR, 200 mm lens, 10 s at f/5.6. The main stars of the Pleiades have been inked in with *Photoshop* to compensate for the inevitable loss of quality in book production. This photo was taken 1 h 5 min after Fig. 6.12. The exposure time had to be increased because we are now looking very close to the horizon, through a lot of cloud and murk. The stars are now noticeably linear. There was not time to set up a driven mount – it was necessary to travel to a place with a good view of the appropriate horizon. Notice that the Moon is now occluding Merope, one of the 'seven sisters' of the Pleiades. Presumably the overhead electric cable looks so bright because it is illuminated by streetlight (Image by the author)

You can obviously see more with larger binoculars than your typical 10×50 models. The 80×20 binoculars, especially if mounted on a tripod, will show you mountains.

You will get a good view with all but the worst telescopes. Do put the telescope on a mount, even if it is only a very simple one.

At magnifications above about 10×, you can pick out and identify the 'seas.' You need about 25× to be able to work out where the Apollo Moon shots landed, although practically no telescope on Earth will show you the spacecraft. The lunar-orbiting probes currently up there can only just spot them. At 100×, you can easily make out individual mountains, although the image will be very shimmery, especially if the Moon is close to the horizon.

The place to look for detail is always near the terminator, the sunrise or sunset line that divides the illuminated part of the Moon from the dark part. Shadows are longer here, which delineates the mountains and the craters.

Telescopes and the Moon

Now let us discuss photographs taken with a telescope. The best way to take such photographs is with a webcam, as described in the photography chapter. If the webcam is in prime focus, the Moon is so big that you have to take several shots and use *Photoshop* or a similar package to merge them into a panorama. This way you get fantastic detail. You can do this with *Photoshop Elements,* but it was a lot easier with a now obsolete package called *Microsoft Digital Image Suite 2006.*

The technique to take a picture like Fig. 6.14 is to shoot about 20 s with a webcam at 5 fps, and stack in *Registax* using a single point to overlap, and sharpen only

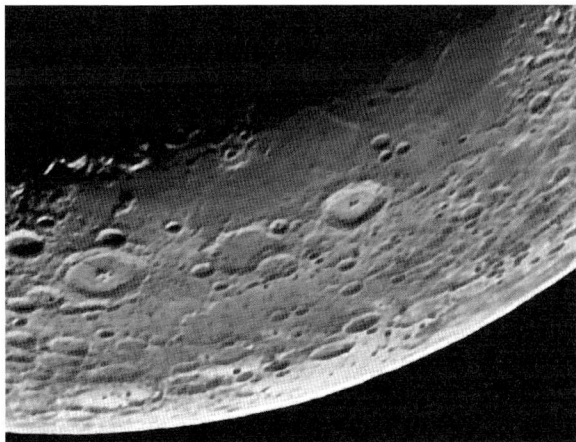

Fig. 6.14 The Moon, April 24, 2007 c. 20:50 UT. The two biggest craters, the ones with central uplands, are Petavius (*left* or south) and Langrenus (*right* or north) (Image by the author)

Fig. 6.15 The Moon, April 24, 2007, c. 20:50 UT (Image by the author)

very lightly with the wavelet facility. The webcam is at prime focus in your telescope, so this is not particularly high magnification. With a Barlow lens, the atmospheric shimmer would also be magnified, and more aggressive post-processing would be needed.

In order to make a panorama like Fig. 6.13 out of pictures, like Fig. 6.14, it is best to forget about color and change the pictures to monochrome. This deals with the problem of reflected streetlight. If you really want a color image, you will have to disable any automatic color correction in your webcam. Even then, if the Moon is close to the horizon, you may find that it appears to redden as you shoot the frames of your composite.

A montage of the full Moon takes about 50 shots. It is a good idea to shoot the whole surface twice to avoid ruining your composite with gaps. For the same reason, there should be plenty of overlap between frames.

Your pictures of very crescent Moons will be less sharp because the Moon is inevitably not very high in the sky unless you photograph this phenomenon on a summer day.

To make montages such as Figs. 6.15, 6.16 and 6.18, you have to do something to add the black background. You partly do this by setting the background color of the montage to be black, but partly also you may have to take the software's 'pen' or 'paintbrush' and do some manual coloring. You also have to manipulate the brightness and contrast of the individual fames and the final composite picture somewhat to achieve a pleasing result.

Notice in each of Figs. 6.15, 6.16 and 6.18 that you can see altitude detail much better near the terminator, because the shadows are longer.

Figure 6.16, the nearly full Moon, shows how at that phase, instead of seeing shadows and relief, you see instead a lot of rays of ejected material (ejecta) from the larger craters. For a long time it was debated whether the craters were of volcanic or meteorite-impact origin. There seems to be little doubt today [44] that they

Fig. 6.16 Nearly full Moon, July 27, 2007, c.22:50 UT. North is *up*, south is *down* (Image by the author)

are mostly impact scars. The 'seas' are remarkably circular. They have quite flat plains, strongly suggesting that they are impact basins that flooded with lava, which subsequently solidified. Comparing Fig. 6.11 with Fig. 6.16 begs the question: why are the 'seas' predominantly on the Earth-facing side of the Moon? The Moon's crust seems to be thinner on this side [45], suggesting that a large impactor is more likely to penetrate to the mantle, which was once molten.

The most visible of the ejecta in Fig. 6.16 come from two craters: Tycho, in the southern highlands, and Copernicus, in a 'sea' (Oceanus Procellarum or Ocean of Storms) in the equatorial region, to the left (west) of center. Both these are relatively late impacts, because they have left their debris on top of other features. The seas were formed before Copernicus, and the mountains before Tycho. It is tempting to see this as evidence that, as the Solar System evolved and planetesimals were pulled together to form planets, the big ones coalesced and impacted before the smaller ones.

Copernicus is shown in more detail in Fig. 6.17. This is in fact one of the individual pictures that make up Fig. 6.18. The mountain range going past Copernicus and Eratosthenes, the Lunar Apennines, is partly submerged by the floor of the Ocean of Storms.

Fig. 6.17 Craters Copernicus (*bottom center*) and Eratosthenes (slightly *right of center*, next to arc of mountains) August 5, 2007, 03:54 UT (Image by the author)

Fig. 6.18 The Moon August 5, 2007, 04–17 UT give or take 25 min (Image by the author)

Impacts as the Dominant Feature of the Moon's Surface

Even the very formation of the Moon could well have been an impact phenomenon: It has been postulated that a collision between two smaller planets produced both Earth and some orbiting ejected debris that later coalesced under its own gravity to form the Moon [52]. Of course, in this scenario, a lot of debris would have escaped altogether, and a lot would have fallen back to Earth. As this theory has been refined over the last decade or so, it has been able to address most of the earlier objections that were raised in regard to it [53]. An alternative hypothesis of capture of the Moon by Earth does not look plausible because it does not explain why the oxygen isotope composition of the bodies is so similar when we know that oxygen isotope ratios are characteristic for each planet [54]. The Moon could not have flown off Earth or co-formed with it because in such scenarios, the Moon would have to orbit Earth's equator. In fact, it orbits much closer to the plane of the ecliptic than to the plane of Earth's equator, in stark contrast to the moons of the gas giants [55].

You can observe this for yourself. The Moon is always near the ecliptic, within 5° or 6°. There is a topocentric component to its position relative to the stars; it is near enough so that its location depends on where on Earth you are. If the Moon orbited around Earth's equator, it would follow a very different path, passing through the middle of such constellations as Orion, Ophiucus and Aquila.

Contrast this with Saturn's moons. Their paths are shown in Figs. 7.45 through 7.49, all photos taken in 2009 when Saturn's rings were edge-on. With Jupiter, you also have to choose an equinox year to see that the moons orbit around the equator of the planet. Jupiter's axis is not strongly tilted, so the effect is not as obvious as for Saturn (Fig. 7.36). It is most obvious of all for the very tilted Uranus, but you would need a powerful telescope to see its moons.

Earth's Moon is also much bigger relative to its 'parent' planet than the gas giants' moons. It is therefore a somewhat different animal from the moons of the gas giants.

Understanding Impact Phenomena

One of the most remarkable facts concerning what we see on the Moon is that they are all fairly round. Even the very largest objects, the 'seas,' tend to be round. This is shown in Fig. 6.19.

We will see that the reason for the roundness of these features is that they are impact scars. Most of them have since filled with molten lava, which of course subsequently solidified. This is why they have flat surfaces. Galileo, of course, mistook them for seas. This is understandable, since his telescopes were not a patch on the ones owned by most amateur astronomers today. He also did not have the benefit of photographs of Earth taken by Apollo astronauts to show him what seas really look like from a quarter of a million miles away. We still call them seas and oceans even though we know better.

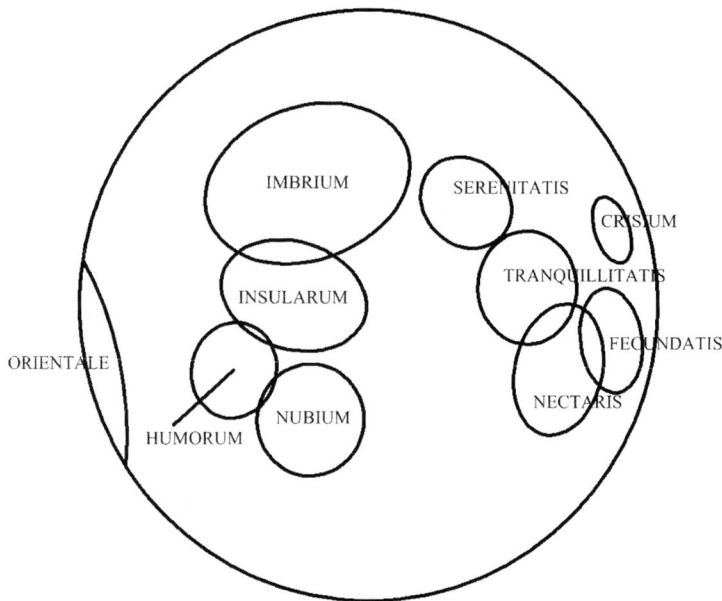

Fig. 6.19 The lunar seas are approximately round or composed of round or rounded regions (Image by the author)

Some club speakers who wish to appear educated use the Latin names common in Galileo's time, but with embarrassing frequency they get the words wrong. The Latin for 'sea' is mare, pronounced 'MAH-ray.' The Latin plural, 'seas,' is *maria*, pronounced like the girl's name with the same spelling. Thus it is correct to say, "This mare is called Mare Crisium." It is also correct to say, "These maria have flat surfaces." It is, however, quite wrong to say "These mare have flat surfaces." Follow this advice and you will be well on your way to bluffing your way through Latin.

As well as maria, one of the first things you will notice if you look at the Moon through binoculars or a telescope is craters. With a 6-in. or better telescope, you will begin to notice that craters come in all sizes, but not all shapes. They are almost without exception remarkably circular. The bigger ones have central features such as mountains.

Let us begin by describing smaller craters. You may need a larger telescope to see these. They have no central feature: they are bowl shaped and are known in the jargon as simple craters.

A typical simple crater on the Moon is shown in Fig. 6.20. It is bowl-shaped. Its rim is raised above the surrounding surface. Its center is below the surrounding surface. There is ejected material, or ejecta, deposited around the crater.

Could you see such a crater with amateur observing equipment? Yes, just! Fig. 6.21 is a 640×480 picture taken at prime focus with an 8-in. f/6 Newtonian telescope. No detail can be seen. But any such craters you see are likely to be simple craters.

Fig. 6.20 A simple crater, 4 miles across. This crater is called Moltke. It was photographed from Apollo 10 (Image courtesy of NASA, photo AS10-29-4324)

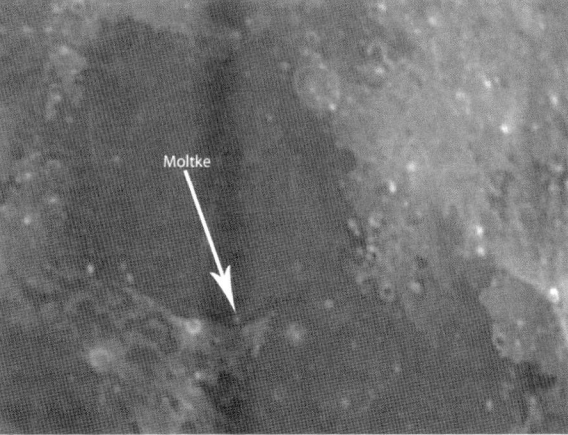

Fig. 6.21 Crater Moltke, 4 miles across. It is on the southern edge of Mare Tranquilitatis, the Sea of Tranquillity. Using an 8-in. f/6 Newtonian, prime focus, Philips SPC900NC webcam, processed in Registax (Image by the author)

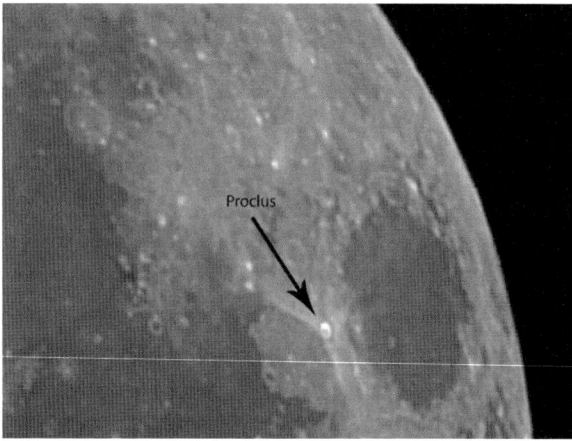

Fig. 6.22 Crater Proclus, 17 miles across. It is on just to the east of Mare Crisium, the Sea of Crises. 8-in. f/6 Newtonian, prime focus, Philips SPC900NC webcam, processed in Registax (Image by the author)

Another interesting simple crater is Proclus, in the highland region just to the west (i.e., left when north is up) of Mare Crisium. An amateur photograph is shown in Fig. 6.22 and an Apollo photograph is shown in Fig. 6.23. This feature is larger, and you can see in Fig. 6.22 that it is a crater. You can also see the ejecta. These are not uniformly distributed about the crater. There is a direction with no ejecta to the south-west. This is because the crater was formed by an oblique impact. The meteorite must have been moving northeast.

Yet curiously, the crater is round, not elliptical. How can this be? We will return to this point below.

It is quite likely that the surrounding smaller craters were made by impacts from the ejecta of this crater.

Until the 1960s, the question of whether lunar craters are volcanic or made by impacts was unanswered. A convincing answer to this question was supplied by the study of terrestrial craters.

There is a well-known crater in Arizona, variously known as Meteor Crater and Barrington Crater (Fig. 6.24).

The first person to hypothesize that this crater was of meteoric origin was Barrington, but he could not prove it. He hoped to find the precious metals of the meteorite buried under the surface and make his million by mining and selling them. Alas he failed, although he did later become a millionaire.

It fell to a very imaginative and energetic geologist called Eugene Shoemaker to prove that this crater was made by a meteorite. He found a mineral below the bowl of the crater called coesite [56]. This is a crystalline form of silicon dioxide that can only be made at absolutely tremendous temperatures and pressures. You just don't find it at volcanic sites. Volcanic activity is not violent enough to create it.

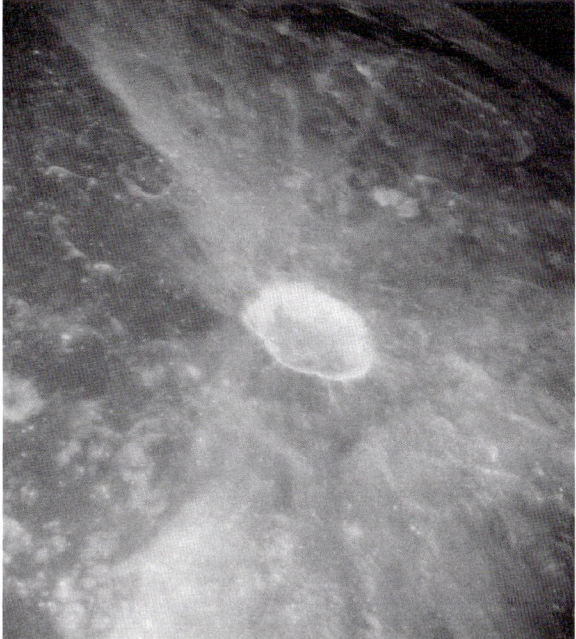

Fig. 6.23 Proclus from Apollo 15. This is still a simple crater, with no central feature (Image courtesy of NASA)

Fig. 6.24 Meteor Crater or Barrington Crater in Arizona. This is also a simple crater. There is no central mountain or other feature (Image courtesy of Shane Torgerson, reproduced under a GNU license)

Fig. 6.25 The Teapot-Ess nuclear bomb test (Image courtesy of the National Nuclear Security Administration Nevada Site Office Photo Library)

Shoemaker had access to geological data from the craters formed by nuclear bomb tests, of which there were alarmingly many in Nevada until the Test Ban Treaty of 1963. People used to come out of the casinos at Las Vegas to watch them. Everyone knew when they would happen because the airport there was closed on test days. In particular, Shoemaker knew that there was coesite in the crater dug by the so-called Teapot-Ess nuclear bomb test (Fig. 6.25).

The volcano hypothesis was thus disproved for Meteor Crater. The only remaining plausible formation mechanism was a very hard meteorite impact.

Shoemaker was heavily involved in the Apollo Moon landing project. Unfortunately he failed the NASA medical fitness tests, or he might have become an astronaut. As it was, he was one of the principal geological trainers of the moonwalkers. This is the same Shoemaker who was one of the co-discoverers of the celebrated comet Shoemaker-Levy 9, which fell onto Jupiter, although his wife Carolyn Shoemaker played a greater part in that discovery than he did. The story of this comet is told in the next chapter. Anyway, thus it was that the Apollo astronauts were trained in crater geology by the world's number one expert on the subject, to the considerable benefit of lunar science.

Fig. 6.26 Two examples of complex craters that are easily visible through binoculars or a modest telescope: Copernicus and Tycho. Copernicus is about 60 miles across; in other words it is the size of a large British county or a mid-sized American one. This shows how much smaller the Moon is than Earth. A county on Earth would be almost invisible on a photo of the whole planet. Note also the ejecta from both craters. Those materials from Tycho go right around the Moon. Some of them probably escaped altogether. The photograph is the same one as in Fig. 6.16 (Image by the author)

The case for the impact theory of crater formation has been greatly strengthened by subsequent study of craters throughout the Solar System, as we shall see. First, however, let us look at craters that do not fit the above description of a simple crater. They are unsurprisingly called complex craters. The two most easily seen examples are shown in Fig. 6.26.

To work out quantities like the floor depth and the height of the central mountains from amateur photos is certainly possible, but it does require some skill in trigonometry and geometry. Fortunately for those who are uncomfortable with such calculations, there are superb NASA photographs of the Moon, many of which are on the Internet. Figure 6.27 is one such image. It shows Copernicus from an angle. We can see that the central floor is flat and below the surrounding lunar surface. We can also see that there are several central peaks, none of which extends as high as the crater rim. Finally, we can see that the inside of the crater rim is terraced.

The flatness of the floor of Copernicus and similar craters is strong evidence that the impact meted the material in the crater floor. In the liquid state, this rock settled into a flat plane – or, more accurately, part of a spherical surface with the same radius of curvature as the Moon. The curvature of the Moon's surface is not

Fig. 6.27 Copernicus photographed by NASA's Lunar Reconnaissance Orbiter (Image courtesy of NASA)

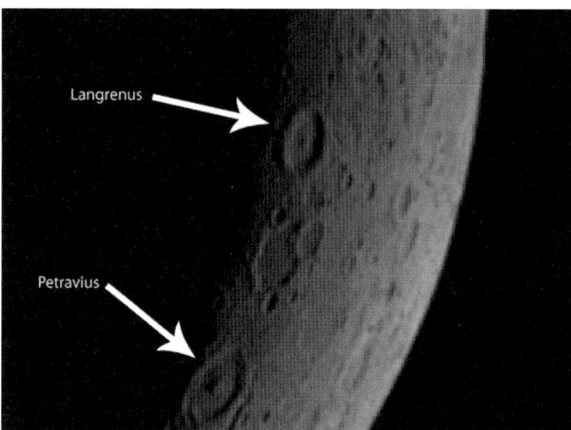

Fig. 6.28 Two complex craters you can easily observe, Langrenus and Petravius. Notice how the flat floor of the largest crater in between these two is flat: any central feature has been covered by solidified molten material. Using an 8-in. f/6 Newtonian, prime focus, Philips SPC900NC webcam, processed in Registax. May 19, 2007, 22:27 UT (Image by the author)

significant for a crater the size of Copernicus, but it is significant for craters hundreds of miles across.

Two other complex craters you can easily observe and photograph are Langrenus and Petravus. They are visible on a waxing Moon, in the southern hemisphere. Figure 6.28 shows these 3 days after a new Moon. You can therefore conveniently

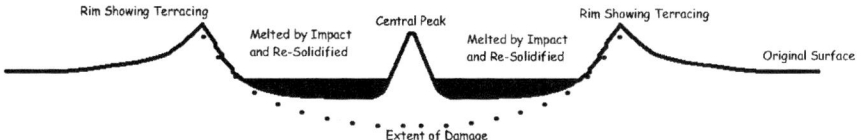

Fig. 6.29 The typical features of a complex crater of modest dimensions. There is a flat base below the original surface layer, a central peak, that would collapse in a larger crater, and a raised rim showing terracing on the inside (Image by the author)

observe them in the evening. Because they are towards the edge of the visible Moon, it is possible to see some of their structure without having to check against NASA photographs. Langrenus is 82 miles across and Petravius is 110 miles across. Not only is Petravius larger, but its central feature is more complex (Fig. 6.29).

What is the origin of this strange feature? It seems odd that a meteorite would blast out a hole all except the bit in the middle.

In fact the central feature formed late in the impact. Two hypotheses have been proposed. The first can be summarized as the elastic rebound model and the second as the slumping model.

The elastic rebound concept [57] imagines that the impacting meteorite pushes the lunar surface down, and although much of it is blasted out of the crater, enough remains to rebound, resulting in the central peak.

The alternative hypothesis is that after the crater has been blasted out, the resultant "transient" crater is unstable, and it rapidly slumps into a new shape [58].

There is another twist to the tale. If the complex crater is really big, the first thing that happens is that the central mountain is no longer a single feature. It can be a ring instead. The nicest example of this is unfortunately not visible from Earth, being on the far side of the Moon. This is Schrödinger Crater, whose diameter is 193 miles (Fig. 6.30).

If the crater is even bigger, there are multiple rings. The classic example of this is the Orientale Basin, which is partly visible from Earth (Fig. 6.19, lower left). It is shown in Fig. 6.31 and is 576 miles across. The impact that produced it must have been quite a whack. Because we see it so obliquely from Earth, it was not recognized for what it is until 1962. Interestingly one of the two people who recognized the multi-ring structure was the Gerard Kuiper, after whom the outer Solar System belt is named [59].

Once Eugene Shoemaker had established the meteoric origin of terrestrial craters, it made sense to study craters on Earth to learn what we could about those on the Moon. The opportunity to drill core samples and otherwise investigate craters on Earth is much greater than that afforded by half a dozen manned and a few unmanned expeditions to the Moon. Laboratory experiments were also performed.

In particular, two groups' experimental work is widely cited in the technical literature: B. A. Ivanov and co-workers [60] and Peter H. Schultz and co-workers [61].

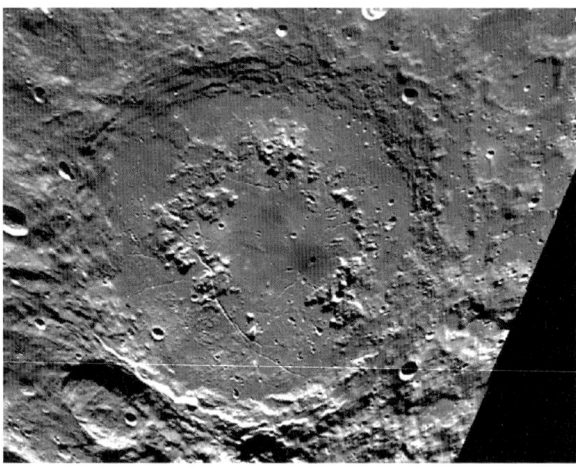

Fig. 6.30 Crater Schrödinger on the far side of the Moon, photographed by the NASA Clementine unmanned lunar orbiter in the mid-1990s (Image courtesy of NASA)

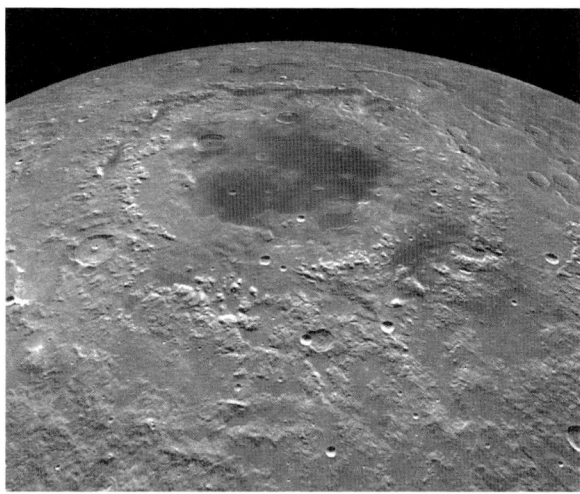

Fig. 6.31 The Orientale impact basin (Image courtesy of NASA)

These groups used guns to fire small pellets into targets of various materials to investigate impact behavior. The term 'gun' may be a little misleading. These guns are enormous and fire their little projectiles more than ten times as fast as a typical speeding bullet – at up to 16,000 miles per hour. The so-called vertical gun at the NASA Ames Research Laboratory is three stories high when in the vertical position. It does not have to fire vertically. It can fire at any angle between horizontal

Fig. 6.32 The NASA Ames Vertical Gun Range for studying crater formation in laboratory conditions (Image courtesy of NASA)

and vertical. The target is placed in a large walk-in chamber strong enough to catch and contain the impact debris. This instrument is shown in Fig. 6.32.

A photo sequence of an impact event created in this instrument is shown in Fig. 6.33. At first we see heat effects as the impactor vaporizes. Then as the crater is dug out, a moving 'curtain wall' of ejecta is created. The initial asymmetry has now disappeared. The ejecta are sent out uniformly in all directions.

Even in a laboratory, a non-vertical impact creates a circular crater.

Figure 6.34 shows the moment of impact. Stress information, in other words, information that the impactor and surface use to react to the collision, travels at the speed of sound in the material. In solids this is typically about a mile per second. The impactor is traveling much faster than that.

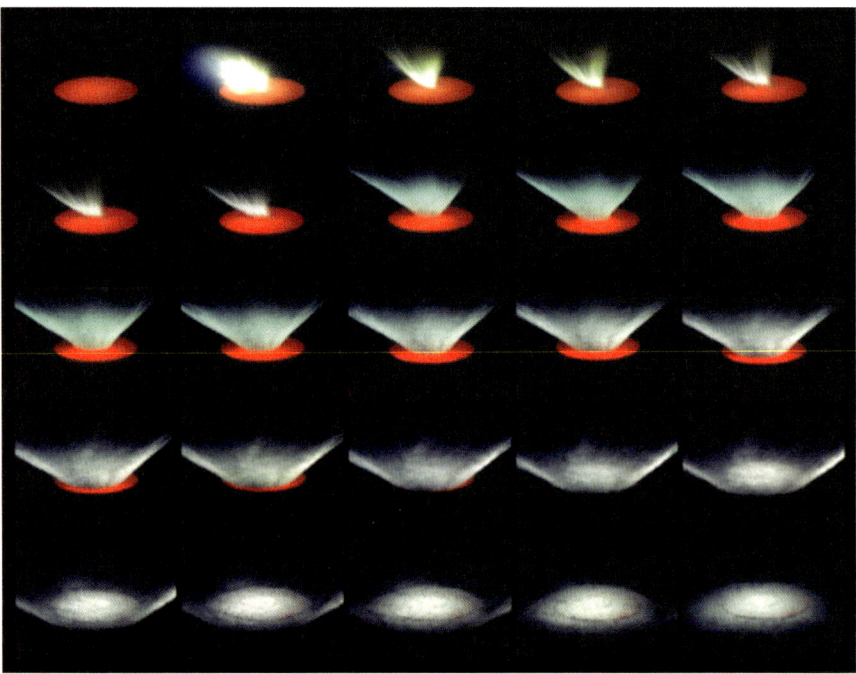

Fig. 6.33 A sequence of photos of a crater created using the NASA Ames Vertical Gun Range for studying crater formation in laboratory conditions (Image courtesy of NASA)

The back of the impactor therefore does not 'know' that an impact has happened. It continues to travel into the Moon as if nothing had happened. The front of the projectile does 'know' it has hit something, and is slowing down, getting damaged, heating up, melting and vaporizing. Therefore the impactor becomes very compressed.

Similar things happen to the surface. It is impacted so fast that stress information cannot reach the rest of the lunar surface. The initial damage is therefore very localized. Material is rapidly ejected from the impact site.

By the time a shock wave is set up, material is already leaving the impact site. This shock wave is like a sonic boom created by a supersonic aircraft, only it is traveling through solid, not air.

This shock wave carries off the remaining impact energy. There is also a shock wave in the impactor. Behind both shock waves, which are compressive waves, there must be a rarefaction, i.e., a region where the material is in tension, the opposite of being in compression, because material that was there has moved into the shock wave.

The rarefaction is what does most of the damage, because most rocky materials are weaker in tension than compression. The meteorite explodes with force

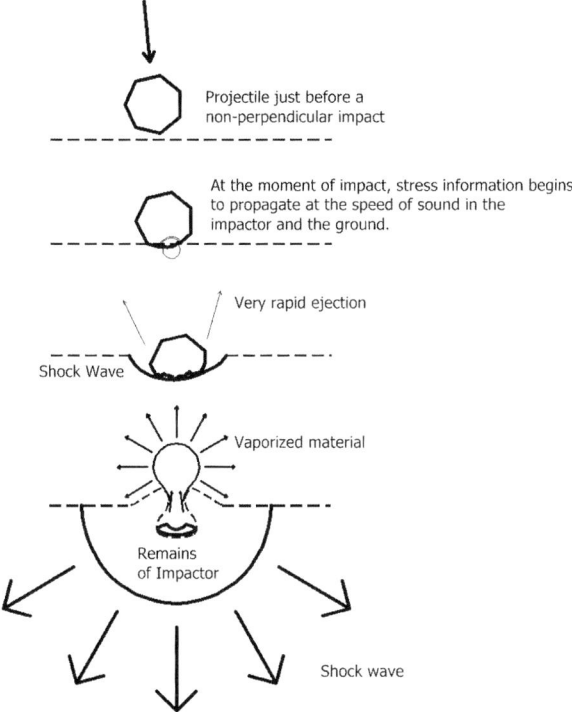

Fig. 6.34 The moment of impact, an event that lasts of the order of seconds for an impactor over a mile across (Image by the author)

comparable to a nuclear weapon, and the crater is excavated. A 'curtain' of excavation propagates through the crater, because it is the rarefaction behind the traveling shock wave that does the digging to excavate the material. Since this shock wave propagates at the speed of sound in the surface material, it propagates just as fast in all directions, irrespective of whether the impact was vertical or diagonal.

This is what makes craters round (Fig. 6.35).

As the shock wave propagates it occupies more space. Its energy becomes diluted. Eventually it is no longer capable of excavating the crater.

We now have a fully dug-out crater. This is called the transient crater, because it is unstable and will collapse.

In Fig. 6.36, the unstable material in the transient crater material is shown hatched.

This material now slumps. The solid black lines in Fig. 6.36 show the final crater shapes. In case (a), a simple crater, the material simply slumps and partially fills the crater bowl. The final apparent crater is a little misleading, because the base material is not quite like the virgin lunar surface material. It is filled with what the geologists call breccia, broken up rock in which the little pieces tend to cement the larger pieces together. Figure 6.37 shows an example of breccia.

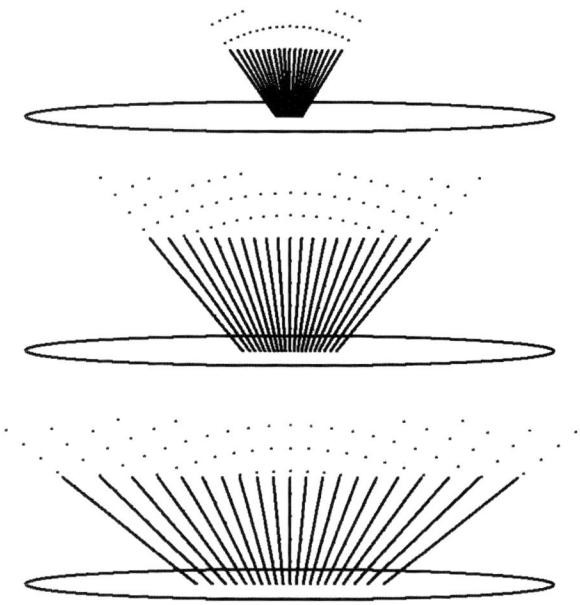

Fig. 6.35 The 'curtain' of material ejected from the crater by the rarefaction immediately behind the shock wave (Image by the author)

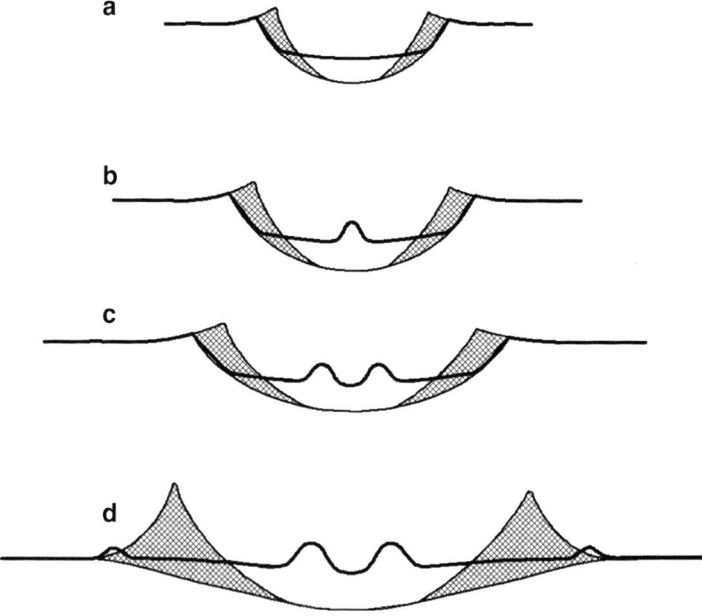

Fig. 6.36 Transient and final craters for (**a**) a simple crater, (**b**) a complex crater with a central mountain, (**c**) a complex crater with a central ring and (**d**) a multi-ring basin. In each case the transient crater material is shown hatched, and the final crater is shown as a thick *black line* (Image by the author)

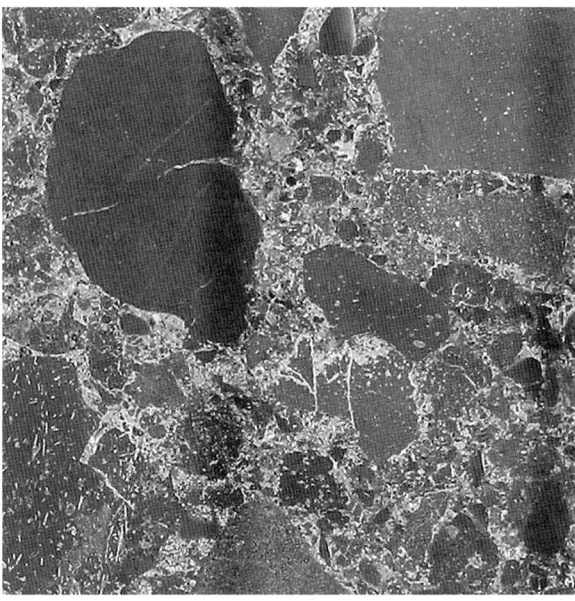

Fig. 6.37 An example of breccia. From the Canary Islands (Illustration courtesy of Siim Sepp, released under a GNU license)

In case (b) Fig. 6.36, the slumping of the transient crater occurs so fast that the slumping material meets in the middle, still going fast enough that it forms a mound.

This is our first statement favoring the slumping hypothesis rather than the rebound hypothesis. In fact Ivanov, who did so much to test it, does not favor the rebound hypothesis [62]. He and his co-worker found that there simply is not enough rebound to form central mountains.

The slumping hypothesis is not without a key complication. It assumes that the slumping material has a particular form. For the computer simulations to work, the breccia in the walls of the transient crater has to behave like a Bingham plastic; and it must be "acoustically fluidized" to have the right properties.

What, you ask, does all this mean? We actually all know what a Bingham plastic is like – we just don't realize it. It is a solid if you leave it in peace, but if you apply a shearing force to it, it turns into a liquid. A good example is toothpaste, which will sit like a solid quite happily on your toothbrush, but turns to liquid once you rub the brush against your teeth. That is how the breccia in the transient crater are hypothesized to behave.

Why would rock behave like that? Melosh and Ivanov [63] postulate that the crater is still ringing from the impact, and that this acoustic ringing loosens the rock enough to make it into a Bingham plastic. It has to be said that there is no direct evidence for this hypothesis. There is only indirect evidence. It is well known in

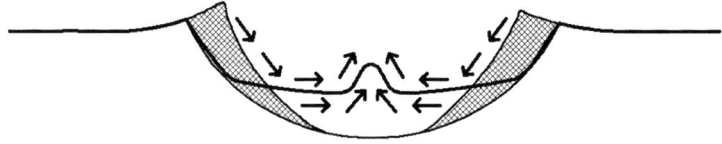

Fig. 6.38 Slumping to form a central mountain in a complex crater (Image by the author)

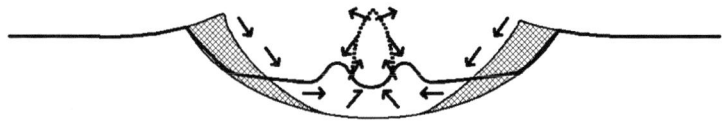

Fig. 6.39 Slumping to form a central ring in a complex crater. The central mountain is now too big to be stable; and it collapses into a ring (Image by the author)

geological circles that soil can liquefy after earthquakes, giving rise to a lot of damage, although this is a slightly different phenomenon. In this case the water from the water table is also shaken to the surface. There is no liquid water to do this on the Moon. Much of the literature on the subject is authored by Melosh, Ivanov or both, but independent support for this acoustic fluidization hypothesis comes from Osinski and Pierazzo [64]. These authors argue that although the idea of acoustic fluidization is controversial, Melosh and Ivanov need only one set of material properties to explain the known crater morphology. The theory would be less likely to be valid if the rock properties had to change for every impact analyzed. That would look like making it up as you go along, which is not a highly recommended scientific procedure.

Slumping to produce a central mountain in the immediate aftermath of the impact is illustrated in Fig. 6.38.

If the central mountain made by the slumping grows too big to be stable, it falls back on itself and collapses, giving a central ring instead of a single peak (Fig. 6.39). The central mountain in Copernicus Crater untidily fits this pattern (Fig. 6.27).

Finally the slumping theory can explain multi-ring basins like Orientale (Fig. 6.31). Rings other than the central one form because the transient crater is big enough to slump outside its rim as well as inside it. This is illustrated in Fig. 6.40.

We therefore have a coherent explanation of crater formation. Smaller craters are simple; larger ones are complex, with a central feature that is a mountain unless the crater is very big, when it becomes a ring. The largest craters have multiple rings.

The craters are formed by a three stage process: impact, excavation by shock wave and slumping of the transient crater to form the final crater. The impacts are so hard that melting and vaporization take place. For very large impactors, enough molten rock is produced to give the crater a flat floor when it re-solidifies.

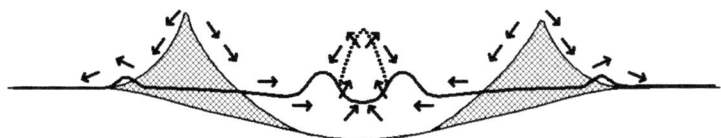

Fig. 6.40 A complex crater with outer rings, also formed by slumping. Once again the central mountain is unstable, and it collapses into a ring. But now, the outer rim of the transient crater is unstable. It also slumps. The formation of the first outer ring shown is by a process not dissimilar to the way in which paint on a vertical wall forms drips if it is applied too thickly. Further outer rings can form by a similar process (Image by the author)

Now, what about the seas? We know that these are not liquid. How else would they contain craters if they were not solid? It is noticeable that that they contain many fewer craters than the highland regions.

We believe that they are super-massive impact basins like Orientale (Fig. 6.40) that filled with lava sometime after the impact. This process buried the existing craters. It neatly explains why the maria have roundish shapes. Since meteorites would be expected to strike maria and highland regions with equal likelihood, we can deduce from the paucity of craters in the maria that they must have filled with lava long after they were formed.

Why did this only happen on the near side of the Moon? It turns out that the far side has a thicker crust than the near side, so that impact damage did not reach down to the molten part of the Moon on the far side, whereas it did on the near side. We know this from seismology studies from the Apollo missions [65] that the lunar crust is thicker on the far side than on the near side. It is generally believed that the Moon's mantle, the layer immediately below the crust, was initially molten, but that it has now solidified [66]. It is not impossible that the cumulative effect of the meteoritic impact at the time of the late heavy bombardment partially re-melted the mantle, enabling the maria to form. The late heavy bombardment is a postulated event to explain the finding from Apollo missions that the cratering seemed to occur 3.8–4.0 billion years ago [67].

The one phenomenon that we have not explained is the characteristic step-like terracing of the inner rims of complex craters, visible in Fig. 6.27. This is not due to slumping. It appears to be a different phenomenon. Melosh suggests that this happens because there is some elastic rebound of the crater floor, enough to cause the formation of the near-vertical cracks between each stage of the terracing [68].

Eclipses

Both Earth and the Moon cast shadows in sunlight, just as you and I do. If the orbits of the Moon and Earth were exactly co-planar, they would pass through each other's shadows every month. The fact that they do not is the most obvious piece of evidence that their orbits are not precisely co-planar.

By coincidence the Moon and the Sun appear to us to be almost exactly the same size. This fact makes total solar eclipses, where the Moon blots out the Sun, quite rare events. They have to pass exactly in front of one another. Because Earth is the much bigger body, each total solar eclipse is only seen along a very narrow line. Most people will see a partial solar eclipse.

Total solar eclipses along a particular line are very rare indeed. They occur much less often than once in a human lifetime.

There was to be an eclipse in the far southwest of England in 1999, on a peninsula called Cornwall. Since the United Kingdom is a smallish country, there is a long gap between solar eclipses visible from any point there. The previous one that had been "visible" (i.e., obscured by clouds) from the UK was a 1,000 miles away in the Shetland Islands almost half a century before, on June 30, 1954.

It does not take a rocket scientist to figure out that if the weather was going to be bad in Cornwall, it would not be possible to follow the good weather along the path of totality. This path would cross the European continent and then head into southern Asia, disappearing into the sunset in India.

The deserts of southern Asia are a long way from Britain. The Internet was then emerging from its infancy and becoming a useful tool, albeit at land-line dial-up speeds. A NASA website gave probabilities of clear weather along the path of totality. It suggested that the best place in Europe was Lake Balatan in Hungary.

A little planning pays off handsomely, as Figs. 6.19, 6.20, 6.21, 6.22, 6.23, 6.24 and 6.25 shows. The best site proved to be 20 miles northwest of Lake Balatan (Figs. 6.41, 6.42, 6.43, 6.44, 6.45, 6.46 and 6.47).

Lunar eclipses are much more common, because Earth, being that much bigger than the Moon, casts a larger shadow; and the Moon, being that much smaller, can be completely within the shadow. The fickle English weather co-operated beautifully on the night of March 3, 2007, and the country was treated to a magnificent view. The method used to take these pictures was an object lesson in how not to do it. The webcam with its lens on was hand held against a 6-in. Newtonian telescope's eyepiece; and the image was processed with *Registax*. It does not matter if you are not Damian Peach, the World's number one amateur Solar System astrophotographer [69]. It is much more important to have a go, and have fun. Figure 6.48 shows what you can photograph, even with dodgy technique.

In Fig. 6.48 you can see that the edge of the shadow is arc-shaped. To the ancient Greeks, this was evidence of the roundness of Earth. It is unfortunate if anyone from the Americas is offended, but the roundness of Earth was not discovered by Christopher Columbus. Thomas Aquinas, writing around the year 1270, took the roundness of Earth as an example of a well-known truth [70]. What Columbus discovered was the New World, not the round world (Fig. 6.49).

Of course, our knowledge of our nearest celestial neighbor has increased out of all recognition since 1959, when the Soviets sent the first space probes there. They, the Americans and the Japanese have all sent orbiters, and the Soviets and Americans have brought back rock samples.

So were the Moon landings faked? The evidence advanced by the advocates of fakery is really easily rebutted. First off, 400,000 people worked on the Apollo

Fig. 6.41 The essentials of low-budget solar eclipse viewing. Sandwiches, liquid refreshment, an inexpensive solar projecting telescope with cardboard tubes and a piece of white paper placed inside a cardboard box to reduce glare. There were seven or eight people from several nations at this spot. They all wanted a look at the telescope image, but they refused our offers of British sandwiches. Why does English food have such a bad reputation? (Image by the author)

space program. Not a single one of them has 'blown the gaffe' on any alleged conspiracy. During the Cold War, the Soviets never claimed that the Apollo landings were faked. Would they really have missed a golden opportunity like that if it had actually existed? There was absolutely no analog of the *Pentagon Papers*, which exposed some embarrassing facts about the Vietnam War. The U.S. flags did not flutter as claimed. The shadows point every which way because the ground was not flat. The trump card of the conspiracy theorists is that there are no stars on the Apollo astronaut's photographs.

Fig. 6.42 A little earlier, this is what the eclipse looked like at 'first contact' (Image by the author)

Fig. 6.43 The Sun gradually disappeared behind the Moon. Even on a sunny August day, by this time the Sun no longer felt warm (Image by the author)

Fig. 6.44 At about the time of Fig. 6.43, the crescent Sun was casting unusual shaped shadows. The fingers of the author's daughter's hand appear to have a small extra digit between them in the picture; and the shadow of the hand is not sharp. The map below shows the path of totality across central Europe. You can just about make out the outlines of Germany, Austria, Switzerland and Italy to the south and the Czech Republic to the north (Image by the author)

Fig. 6.45 Something unexpected. The clouds to the *right* of this image are already in darkness (Image by the author)

Fig. 6.46 Compare this shot to Fig. 6.45. It is getting dark where we are, but the cloud to the right is back in daylight again. The birds and the crickets went into their dusk routine at this point. The temperature dropped noticeably. The thermal convection causing the cloud to bubble up also stopped, and the cloud never recovered. Instead it spread out into a flat sheet. I have heard many tales of how cloud cleared right at the moment of a solar eclipse. I can attest that solar eclipses do affect the weather slightly along the path of totality (Image by the author)

Fig. 6.47 The corona, August 11, 1999, c. 11:00 UT. All I could do with the equipment available to me was to take this hand-held shot. Presumably the funny shape in the middle of the corona is the result of camera shake. The bright splotch below and to the *left* of the corona might be Mercury. Venus was almost at inferior conjunction, and would have looked rather like the rightmost image in Fig. 7.12 (Image by the author)

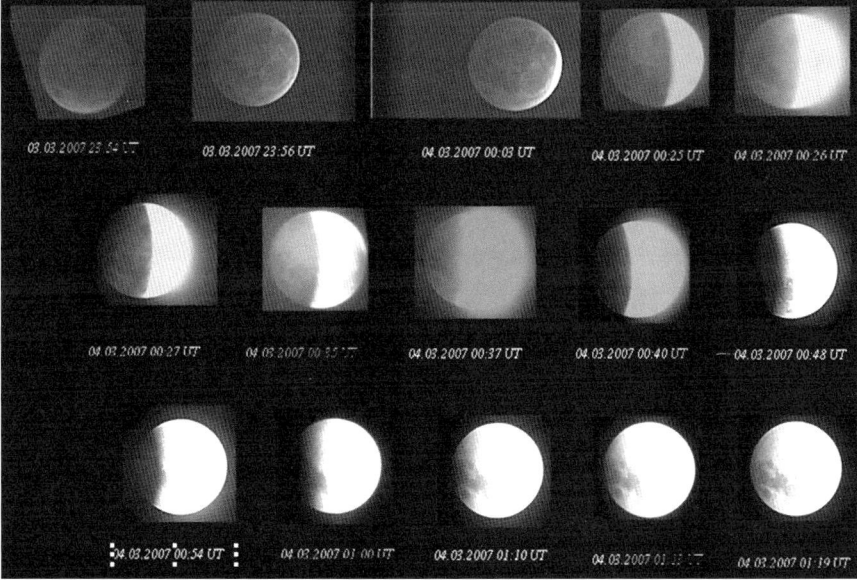

Fig. 6.48 Sequence of images of the total lunar eclipse of March 3–4, 2007. The shaded region has an *orangey-red* color during totality, which is due to the fact that Earth's atmosphere transmits *red* light better than *blue* light, and some of this *red* light illuminates the shadow region. The exact color changes with every lunar eclipse (Image by the author)

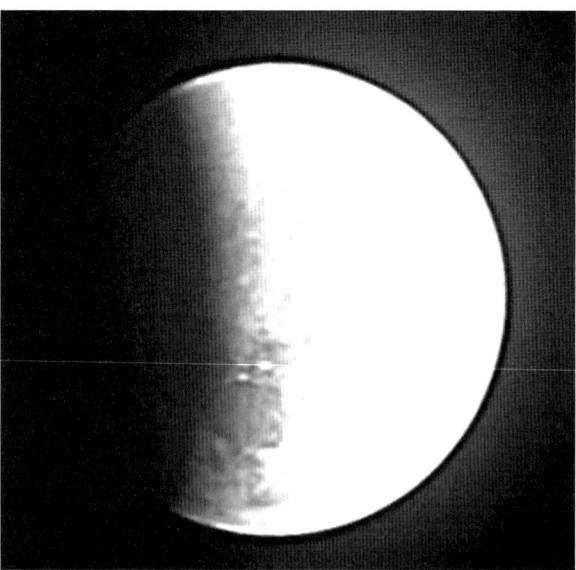

Fig. 6.49 Contrast the edge of the shadow here with the terminator in Fig. 6.18. In Fig. 6.18, the terminator is sharper and is the line along which the shadows are most marked. Here the Moon is full, and there are no shadows cast by mountains along the edge of Earth's shadow. This edge here is also fuzzier than the terminator in Fig. 6.18. March 4, 2007, 00:48 UT (Image by the author)

Well, those of us who observe the heavens will tell you that even when it is a quarter of a million miles away the Moon is bright enough to render all but the brightest stars invisible. One of the most remarkable things about witnessing a total lunar eclipse is that with a full Moon in the sky you can see nearly as many stars as on a moonless night. How much brighter will the Moon seem when it is only three or four feet below the cameras strapped to the astronauts' chests? The exposure times for the photos would have been way below that needed to capture stars.

Finally, you will hear a certain amount of talk of transient lunar phenomena. Unless two or more observers simultaneously photograph a phenomenon, there is no guarantee that it happened. Visual observation, even by several people, is too likely to be distorted by wishful thinking and a desire for glory. People have photographed flashes from meteorite impacts, usually on the unlit side of the Moon. This video footage certainly seems to be real, but otherwise a little skepticism is healthy.

Chapter 7

The Planets: What You Can Realistically Expect to See

We were all brought up on photographs from the Hubble Space Telescope and the various interplanetary space probes. It goes without saying that the space probes had state-of-the-art telescopes and cameras on board, so they sent back some fabulous pictures. Some of them, notably the Cassini probe orbiting Saturn, continue to do so at the time of this writing.

Consequently we have been rather spoiled by these pictures and are apt to expect to reproduce them in our backyards, with equipment costing no more than a few hundred bucks. Mostly it's the last 100 miles traveled by the light from the planets that cause the problems, because of the inconvenient fact that Earth has an atmosphere, and we, rather selfishly, would rather we could breathe.

To climb above that atmosphere is an expensive enterprise. The Hubble Space Telescope cost about $6 billion, plus a European contribution of 600 million Euros according to Wikipedia [71]. It orbits 347 miles above sea level, which is a tiny fraction of the distance to the Moon, let alone the distance to Neptune.

The professionals now observe from the tops of mountains when they don't get Hubble time. The best mountaintop telescopes can now outperform the Hubble, not least by using adaptive optics, whereby the mirrors are made up of many separate hexagonal sub-mirrors that can be pumped in and out several times a second to compensate for atmospheric shimmer [72].

Alas, all this is beyond the budgets of amateurs. But don't let any of this spoil your fun. It is always a pleasure to see the expressions of delighted surprise from people who have never before looked at a planet through a telescope.

Please don't give up in despair if you can't get better pictures than NASA. If anything, it is remarkable what amateurs *can* photograph these days. The improvement in recent years has been considerable.

© Springer Science+Business Media New York 2015
J. Clark, *Viewing and Imaging the Solar System*, The Patrick Moore
Practical Astronomy Series, DOI 10.1007/978-1-4614-5179-2_7

Mercury

Since these two planets are nearer to the Sun than we are, they are never found far from the Sun. You see them around sunrise and sunset.

Mercury, the closest planet to the Sun, is never visible when it is dark. We only see it in the dawn or dusk. It is always close to the horizon. Mercury can move up to 28° from the Sun. When it does, it is always best seen from low latitudes in the Southern Hemisphere – readers in northern Australia or southern Africa may actually see it against a fairly dark sky. In North America and Europe, especially Europe because the populated English-speaking area is above 50°N, further north than Vancouver or Winnipeg, you will not get a good view.

What you can see at the horizon and what you can see 30° above the horizon are very different. You have to look through much more atmosphere near the horizon.

The only photo the author ever took of Mercury is shown in Fig. 7.1. It took quite a lot of planning.

Some much cruder photos were taken to ensure that this really was Mercury, not some other object. They were not very good pictures, but they proved the point. Then an equatorial mount had to be set up somewhere with a good view of the horizon. The next clear morning, the planet was so low in the sky that it was

Fig. 7.1 The author's only photograph of Mercury, taken through a telescope on the morning of November 12, 2007, at 06:56 (8-in. Newtonian, 3.4× Barlow lens, Philips SPC900NC webcam, processed in Registax). No detail is visible. You can see that it is not a star, and that it has a phase. Because the planet was at rooftop height, there was so much atmosphere-induced chromatic aberration that the *red, green* and *blue* images were completely separate from one another (Image by the author)

Fig. 7.2 Mercury and the Moon against a backdrop of the Pleiades, April 26, 2009, 20:35:09 UT. Mercury is towards the bottom of the picture, just to the right of the trees. Image details: Canon EOS100D, 80–200 mm zoom set to 103 mm, 5 s at f/5.6. I tried several exposure times. Eucalyptus trees are not native to England. They are common in the east of the country (Image by the author)

necessary to wait for it to pass behind a chimney. By this time the sky was quite bright. Sunrise was only 17 min away. If you do track Mercury beyond sunrise, please do it carefully. NASA will not point the Hubble Space Telescope at Mercury because of the risk of damage from sunlight. You too should be careful not to damage your eyes or fry your webcam.

The next few pictures show you just how fast Mercury moves through the sky as it plies its orbit. Around May Day 2009, Mercury traveled in front of the Pleiades.

To take the picture in Fig. 7.2 over the back hedge it was necessary to put a DSLR in the top-floor bathroom, which had a window looking to the northeast.

Figure 7.5 particularly shows how difficult Mercury is to photograph, and all of Figs. 7.2, 7.3, 7.4 and 7.5 show you must expect it to move from night to night.

What do we know about this planet? It is actually the least studied of the naked-eye planets because not only is terrestrial observation difficult, but so is extraterrestrial observation. We have mentioned the lack of Hubble pictures. Because this planet moves so fast, we actually cannot fire a rocket fast enough to catch it up. The probe orbiting Mercury at the time of writing, the Messenger probe, had to be gravitationally slingshot from Earth once, Venus twice and Mercury itself

Fig. 7.3 Three nights later, Mercury has moved closer to the Pleiades and eastwards. It is barely above the tree leaves. April 29, 2009, 20:48:49 UT. Image details: Canon EOS100D, 80–200 mm zoom set to 200 mm, 2.5 s at f/5.6. Stars and planet inked in with Photoshop (Image by the author)

Fig. 7.4 By the next night Mercury has moved further east still. April 30, 2009, 20:26:22 UT. Mercury is barely above the tree leaves. Image details: Canon EOS100D, 80–200 mm zoom set to 200 mm, 2.5 s at f/5.6. Stars and planet inked in with Photoshop (Image by the author)

Fig. 7.5 Four evenings later Mercury has moved further east still. May 3, 2009, 21:01:11 UT. Image details: Canon EOS100D, 80–200 mm zoom set to 200 mm, 2.5 s at f/5.6. Stars and planet inked in with Photoshop (Image by the author)

three times. The journey took 7 years [73], when Mercury is about one AU away from us, compared to 7 years to get the Cassini probe about 10 AU from Earth to Saturn [74]. You also have to slow the probe down once it gets there. It has to survive and operate in the presence of very intense solar radiation. Quite a challenge!

We know that Mercury is the densest planet in the Solar System. That means it must be a rocky planet, not a gaseous one. The only way to measure a planet's mass is by observing how its gravity affects other bodies. Mercury has no moons. Its mass could be estimated by studying its slight perturbations of the orbit of Venus. The first accurate measurements came from observation of the first space probe to fly past it, which was *Mariner 10* in the mid-1970s.

Knowledge of the size comes from analysis of its orbit, and measurements from photographs – hopefully better ones than Fig. 7.1. To get the density, you simply divide the mass by the volume.

The four rocky planets are made mainly of iron and silicon, so we can surmise from the greater density that there is more iron and less silicon in Mercury [75]. The conventional wisdom is that this is due to a serious collision early in Mercury's life. How do we know this? The explanation is quite simple. When the planets form, they are largely hot and molten. The iron tends to sink, and the lighter silicon tends to float. If the planet later gets a big whack, it is the silicon nearer the surface that gets knocked off.

Mercury has next to no atmosphere. The nearby Sun would long since have boiled any atmosphere away. Mercury's surface gravity is only 37 % of that on Earth. Gravity is the only thing that keeps a planet's atmosphere there. The planet is pockmarked by craters, like the Moon. We only really know this from extraterrestrial observation. Pre-space-age claims made about the existence of such craters were not borne out by the *Mariner 10* mission. These craters are therefore not something you can expect to see for yourself.

From radar studies it was established in the 1960s that a day on Mercury is exactly two-thirds of a Mercury year [76]. A Mercury year is 88 Earth days. Four Mercury years are therefore 352 of our days, or almost an Earth year.

Because Mercury is closer to the Sun than we are, it will occasionally pass in front of the Sun. These rare events are called transits. In the 10 years following the writing of this book there will be two transits, on May 9, 2016, from 11:12 to 18:42 UT and on November 11, 2019, from 12:35 to 18:04 UT [77]. If the weather cooperates, they should be visible from southern Africa, Europe and North America.

At mid-transit in 2016 in London, the Sun and Mercury will be 45° above the horizon. In Vancouver, they will be 11° above the horizon. In Cape Town, they will be 20° above the horizon. It will be midnight in Sydney, so the transit will not be visible from Australasia [78].

At mid-transit in 2019 in London, the Sun and Mercury will be 11° above the horizon. In Vancouver, it will be sunrise. In Cape Town, they will be 23° above the horizon. It will be 1:20 a.m. in Sydney, so the transit will again not be visible from Australasia [78].

In summary, then, the first challenge with Mercury is to find it – it lives up to its name and moves across the sky very fast. You can only see it at dawn or dusk unless you are somewhere at low latitude in the Southern Hemisphere such as Darwin, Australia, at the time of greatest eastern or western elongation, when it will be reasonably dark by the time Mercury sets. Because it skims the horizon, you should be able to photograph its phases, but not much more.

Venus

Venus is a much more promising target for amateurs. When it is near elongation, even in northerly England, it is above the horizon for a couple of hours after sunset. It is very bright, brighter than any celestial object other than the Sun or Moon. For those who like numerical magnitudes, it has a magnitude between −3.8 and −4.9. As Fig. 7.6 shows, it is not hard to spot Venus in the dawn or dusk.

The phase of Venus when Fig. 7.6 was taken is shown in Fig. 7.7. Patrick Moore reports in one of his books an experiment he did to test whether people's claims to be able to see the phases of Venus by the naked eye were true. He used to be a schoolteacher, and told classes of schoolboys that the crescent was upside down from the way it actually was. He found that very occasionally a boy would challenge this and say that actually it was the other way up – so he actually could see the phase. However, most people cannot [79].

Venus is the most similar planet in size to Earth. It is a bit smaller. Its surface gravity is about 90 % of that on Earth (Table 1.1). When people speculate about colonizing other planets now that we have filled this one up, they cheerfully forget

Fig. 7.6 Venus in the sky taken from my yard on February 21, 2009, around sunset. Exposure time 3.5 s at f/5.6 using my Canon EOS1000D DSLR with a zoom lens set to 18 mm focal length. Venus is the *white dot* left of center, next to the edge of the clouds. Even though the Sun has yet to set, Venus is still clearly visible (Image by the author)

Fig. 7.7 A few minutes before the taking the picture in 7.6; this one was with a webcam through an 8-in. Newtonian telescope. Philips SPC900NC webcam, processed in Registax (Image by the author)

that a person growing up on the Moon (16 % of Earth's surface gravity) or Mars (38 % of Earth's surface gravity) would grow up rather differently from an Earthling. Even aboard the International Space Station, adult astronauts lose bone density. The only place with a solid surface and the right gravity is Venus.

Unfortunately, if General Sherman, who famously said that if he owned hell and Texas, he would live in hell and rent out Texas, had also owned Venus, he'd be renting that out for the same reason as pre-air-conditioning Texas – it would be too jolly hot. It is even hotter than Mercury because its carbon dioxide atmosphere gives rise to a greenhouse effect worse than Al Gore's worst nightmare [80]. If this inconvenient truth were not enough, the atmosphere of Venus is also full of sulfuric acid.

Most of us learned in school chemistry that acids attack metals. We probably did not learn that some of our most useful non-metallic materials, plastics and rubbers, fall apart at these temperatures. You even need special metals: most everyday steels start to lose strength above 200 °C, and many aluminum alloys go soft much above 100 °C. Venus' surface temperature is nearer 460 °C. Transporting our civilization to Venus would be a very tricky proposition.

Then there is the atmosphere. It is not only hot and acidic, but there is also an awful lot more of it than on Earth. The density of our atmosphere at sea level is about 1.2 kg per cubic meter. On Venus it is more like 65 kg per cubic meter – 54 times as dense. The atmospheric pressure is over 90 times as at sea level on Earth. So if we were to try to live there, we would have to keep out heat, pressure and acid over large enough regions to grow our food. Not easy.

Not exactly thoughts of what Venus, as the goddess of love and beauty, is supposed to evoke.

So what can we see of this hell hole? With visible light all you can really expect to see is the phase; views of Venus are pretty black-and-white. You can also watch it change size.

To be crescent, Venus has to be nearer to us than the Sun. For it to be gibbous, it has to be further away than when it is crescent. It therefore appears to be smaller when it is gibbous, as shown in Fig. 7.8.

In ultraviolet light, streaks can be seen in the clouds. You can photograph this with a webcam, although the UV (ultraviolet) filter is expensive for just for one type of photo. The photos NASA takes from space, e.g., Fig. 7.9, are so much better. You can see the streaks, but not much more.

Some people have caused considerable controversy by sketching these clouds, which most people cannot see at all when looking through a telescope. The sketchers have met with incredulity. The explanation is simple. Some people see further into the UV than others. A rough-and-ready test of whether you are one of the minority of UV-seers is to look through a glass window and compare the colors to what you see with the window open so that you are not looking through glass. Glass blocks UV. If you notice that the glass changes colors, this may well mean that you see very near ultraviolet. Those who can see UV light report that the effect is especially noticeable when they look at grass. Unless you live in the desert, the grass-through-glass test will show whether you see near UV.

You can see Venus in daylight. Once you have done that, you can point your telescope at it, and see it quite well. When it is a morning planet, you can align your telescope on it before the sky brightens up, and then follow it into the day more or less indefinitely.

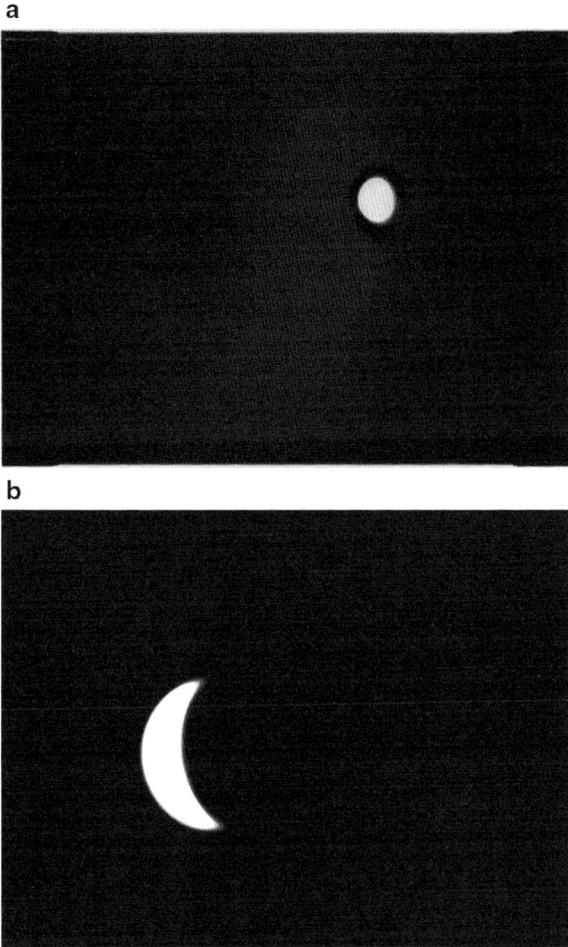

Fig. 7.8 Venus can be gibbous (*left*, January 30, 2008) or crescent (*right*, September 16 2007). 8-in. Newtonian telescope, Philips SPC900NC webcam, processed in Registax (Image by the author)

With practice you can find Venus in the daytime sky with binoculars. This is not easy, but it can be done. The technique is to know where it is relative to the Sun, and use the corner of a building to shade you from the Sun, which means your pupils don't close up quite so much, and it also removes the risk of frying your eyeballs by looking at the Sun through binoculars. If you do not have a handy corner of a building, don't try this – it is too risky.

Anyway, back to the clouds. The clouds surrounding Venus are completely opaque to visible and UV light. We can't see through them at all. We did not even know how much Venus rotates about its axis until the 1960s. Then radar studies

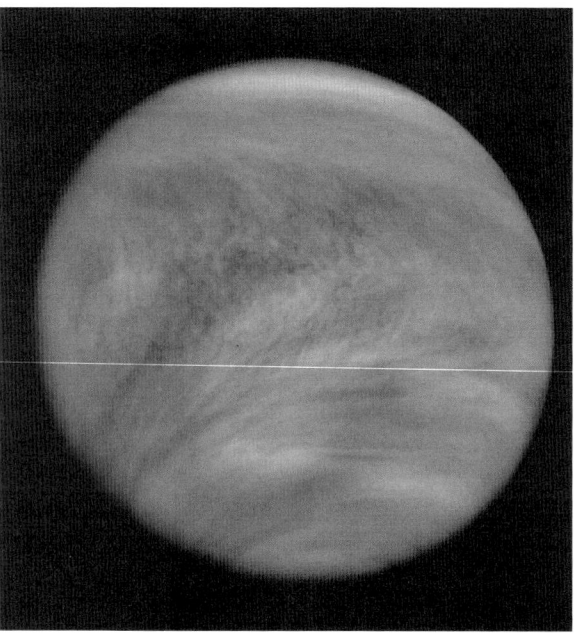

Fig. 7.9 Venus photographed in UV by the Pioneer Venus Orbiter, February 26, 1979 (Image courtesy of NASA)

found that it is one of only two planets in the Solar System that rotate clockwise when viewed from above the north pole of the Sun. This is obviously an imaginary view. No person or space probe has been there. Such rotation is called *retrograde*. The planet rotates very slowly about its axis once every 243 Earth days. A year on Venus is 224.7 Earth days, so it takes more than a year to rotate about its axis. If you work it out, a solar day on Venus would therefore be 116.75 days. Because the rotation is retrograde, the Sun would rise in the west and set in the east, except that you would not see it because of the clouds. Imagine going all that way, further than Florida, and quite a bit nearer the Sun, only to come back with no suntan whatsoever.

Radar was also used in the 1960s to measure the distance to Venus. Radar is electromagnetic radiation in the radio or microwave frequency range. It travels at the speed of light. By timing its journey, the distance can be obtained. This method is the current 'gold standard' for measuring the astronomical unit. Orbital studies will give you the relative distances of the members of the Solar System from the Sun, but not the absolute distances. You need an independent measure of the astronomical unit.

Bouncing radar signals off another planet, millions of miles away, was a whole different game from bouncing radar off enemy aircraft. It was not at all an easy thing to do, and there were many false starts. Eventually, Richard Goldstein of the

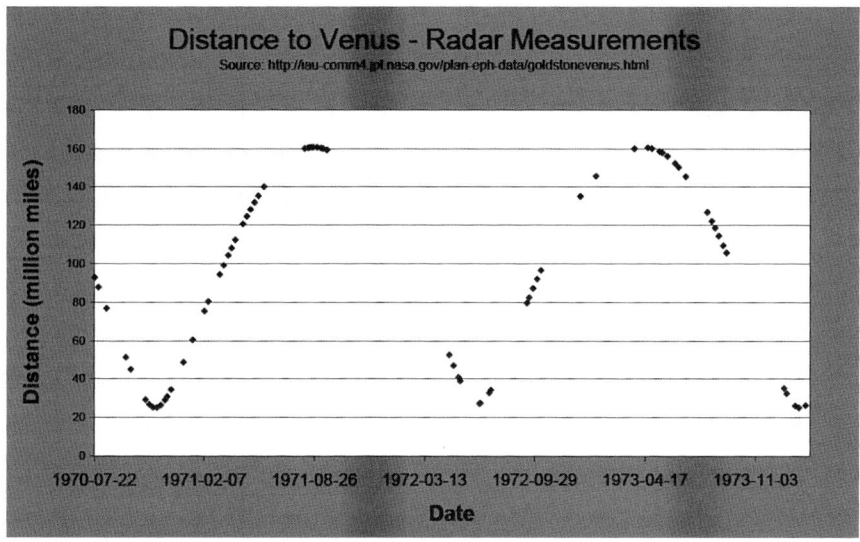

Fig. 7.10 Radar ranging data from NASA's Jet Propulsion Laboratory. The website expresses the data as time delays. An accepted value of the speed of light has been used to convert the data to distances in miles (Image by the author)

Goldstone radar facility run by NASA's Jet Propulsion Laboratory, figured out how to solve this problem He was a Ph.D. student. Anyone with a doctorate will tell you that Ph.D. students are the lowest of low people on the scientific totem pole, treated every bit as roughly as apprentices in any other profession. Goldstein had to put up with much skepticism from his professor, who knew all about other peoples' failures to make this measurement. It says much for him that he had the courage to go ahead.

The JPL has put some of its radar data collected over the years online, although apparently not Goldstein's data. Some of the data are plotted in Fig. 7.10.

It can be seen from Fig. 7.10 that the distance from Earth to Venus varies enormously. The apparent size of Venus will also therefore vary considerably. The photographs in Figs. 7.11 and 7.12, show this.

The video capture software, *K3CCDTools,* converts a picture size in pixels to arc seconds. In Fig. 7.13 the sizes of Venus from the photos in Figs. 7.11 and 7.12 are compared with what published data lead us to expect.

A reasonable conclusion is that the values obtained are much better when Venus is close than when it is far away. The reason for that is simple enough: a slight error in a big image is relatively smaller than a similar error in a small one only occupying a few pixels.

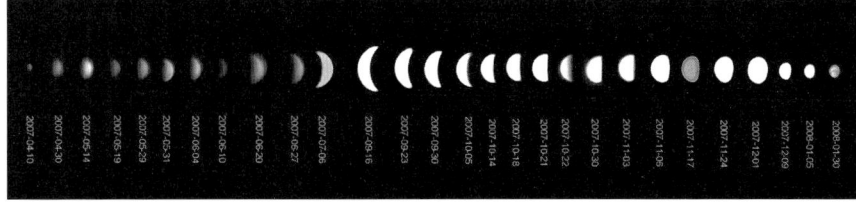

Fig. 7.11 2007–2008 apparition of Venus (Image by the author [81])

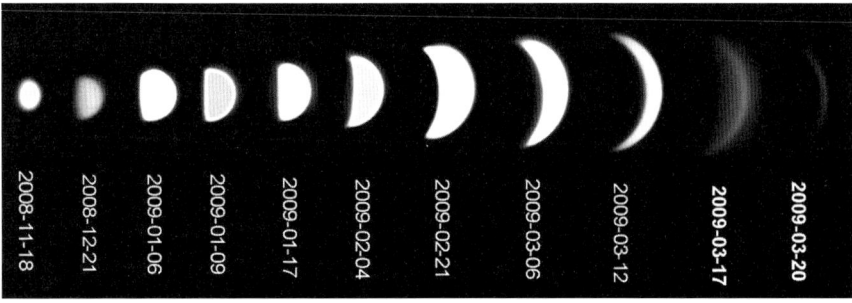

Fig. 7.12 2008–2009 apparition of Venus. It may be noted that the author's photographic technique had improved since the photos in Fig. 7.11 were taken (Image by the author)

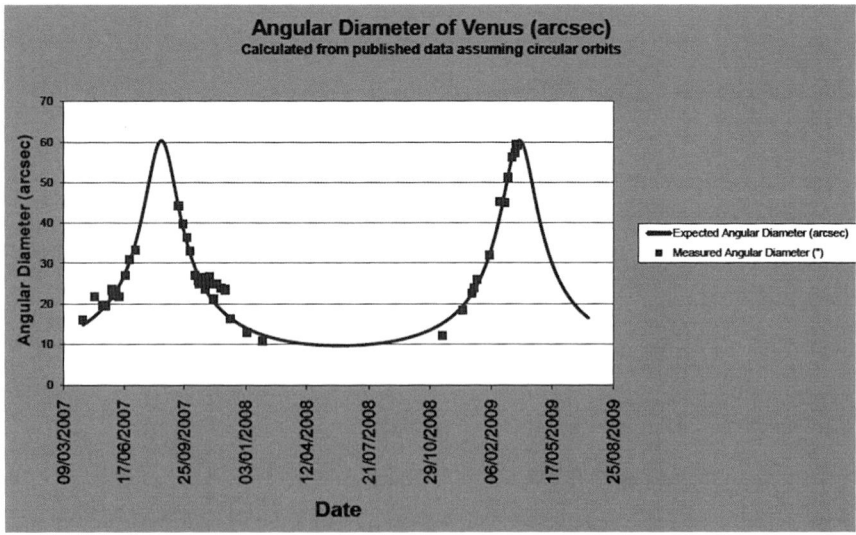

Fig. 7.13 Apparent sizes of Venus measured by the author compared to the expected values, assuming published orbital and size data (Image by the author)

Mars, Every Other Year

Mars takes about 2 years to complete an orbit. The time between oppositions is actually 2 years and 49 days. In between oppositions it can be found, but it is too far away to be a practical observing target for amateur astronomers. It is a small planet – not as small as Mercury but considerably smaller than Earth. It therefore needs to be relatively close to us to appear as a decent-sized object in a telescope.

It has a diameter a little over half of ours, and therefore a volume (proportional to radius cubed) of a little over one eighth that of Earth. It is less dense than Earth on average, and has a mass of a smidgen over one-tenth that of Earth.

It also rotates quite fast. A day on Mars (called a 'sol') is 24 h and 37 min, not a whole lot different than what we experience here (Fig. 7.14).

Might humans settle on Mars? Hmmm. The challenges are a bit less daunting than those related to Venus, but they are still enormous. Neither planet has much of a magnetic field. Our magnetic field protects us from solar radiation, by deflecting

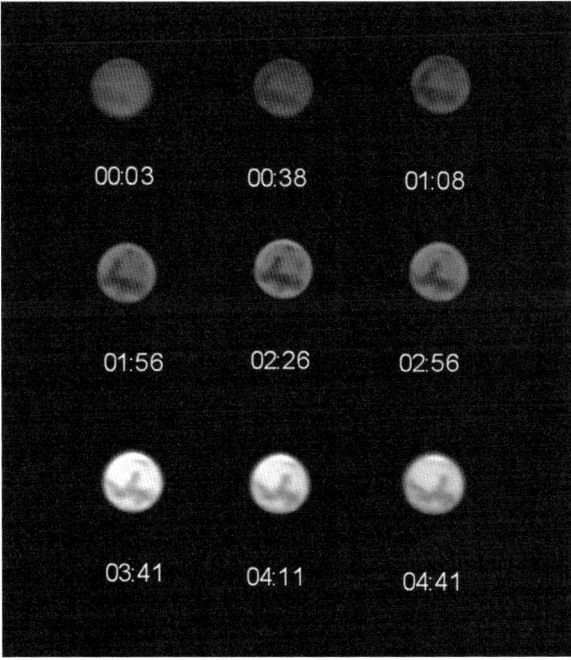

Fig. 7.14 The rotation of Mars during a single night at about the 2007 opposition. The visible north–south (i.e., roughly vertical) feature is known as Syrtis Major (Image by the author)

it around Earth. Obviously on Mars the radiation would be less intense than here or on Venus, but we would not only have to protect ourselves from it but also our livestock and crops. There is only a very thin atmosphere there, barely more than a vacuum. We would have to take our own atmosphere. It is quite a tall order to try to build a canopy over a whole planet. It would be a lot easier to colonize our oceans. The environment there is less hostile than elsewhere. We would also need to do something about the gravity problem. There isn't much there. However, it is not too difficult to make artificial gravity – build your settlements as giant centrifuges.

Overall, though, there is not much cause for optimism. The best we might manage is to put a small colony there in the hope that it would provide a small population to restart life here were there to be a nuclear war, which is perhaps inevitable given humankind's unfailing capacity for being both the cleverest and the stupidest creatures alive.

Mars, being both relatively close and a less hostile place than Venus, is the most thoroughly explored of all the planets. The exploration has of course all been by unmanned probes. As our robotics technology improves, there is less and less scientific incentive to send people to explore the Solar System. It is becoming the age of the armchair astronaut.

The history of terrestrial observation of Mars is a mixture of the excellent and the woeful. The excellence has been in the determination of the orbit and the rotation. Kepler's determination of the orbit of Mars was one of the cornerstones of modern physics, which led Newton to discover the mechanism of the forces determining this orbit [82].

As telescopes gradually improved, more and more of the planet's surface features became visible. The story began in 1877 with an Italian, Giovanni Schiaparelli, claiming to see 'canali,' a word that should be translated into English as 'channels' but got translated as 'canals.' An American with more money than sense, Percival Lowell, took up the cause of these canals, and made a great deal of hay popularizing the idea [83]. The power of wishful thinking is amazing. Scientific articles on Martian biology appeared right up to the arrival of the first space probes [84].

The first flyby took place in 1965. The photographs sent back showed an arid desert pockmarked with craters. Yet still the searchers for life have not given up. A recent idea was that some meteorites from Mars – fragments of meteorite strikes or volcanic eruptions – which had found their way to Earth contained fossils of very primitive life forms. The biologists never really bought this idea. The fossils are a lot smaller than any known and firmly established bacteria. For them to be fossils of living creatures, we will have to discover a whole new class of 'nanobacteria' [85].

The existence of water on Mars is now proved beyond a reasonable doubt [86]. It doesn't exist in large quantities at the surface. Liquid water requires the right temperature range, which depends on the atmospheric pressure [87]. There's hardly any pressure on Mars' surface, and little protection from radiation, so we would not expect to find lakes or seas. The exciting questions are: how much is there and

Fig. 7.15 Mars, September 16 2007. Early in a Mars apparition, Mars is quite some ways away, small, and we see a lot of the night side of the planet – or rather, we see that we can't see it. The north polar ice cap is visible. The *white* region to the south makes one wonder how real this sighting of the pole is. Taken using an 8-in. Newtonian Telescope, Philips SPC900NC webcam, processed in Registax (Image by the author)

where is it? If humans ever go there, it would be an awful lot easier not to have to transport the water they need.

However thin it may be [88], the atmosphere is anything but passive. There is weather on Mars. There are seasons. We amateurs can observe all this. The ice caps at the poles change. The surface features are occasionally obscured by dust storms. [89]

It is much easier to observe these phenomena with a webcam than just by looking. Most people simply see an orange shape through telescopes. Filters help – they do increase the contrast on Mars. You might even think about a neutral filter when Mars is close to opposition, because Mars is a bright object. But the best way to demonstrate Mars to non-expert observers is to hook up a webcam and a laptop to the telescope. Even live, before you have used *Registax* to improve the picture, you see more than just by looking through the 'scope.

Figures 7.15, 7.16, 7.17, 7.18, 7.19 and 7.20 show a series of pictures taken during the 2007–2008 apparition of Mars. Part of the fun of observing Mars is that it does keep changing. Venus does, too, but you see so much more of Mars even though it is smaller and on average further away.

Mars has two tiny moons, Phobos and Deimos. Neither is much more than a giant boulder. They may well be captured asteroids. Asteroids are more common near the orbital path of Mars than near that of Earth. They are about 19th magnitude, so you will need a powerful telescope to detect them [90].

Fig. 7.16 Mars, September 30, 2007. This time there is no mistaking the polar ice cap. Other continent-sized features can be seen. 8-in. Newtonian Telescope, Philips SPC900NC webcam, processed in Registax (Image by the author)

Fig. 7.17 Mars, December 5, 2007. The surface features are different from those seen before. The phase is now much closer to full. Taken using an 8-in. Newtonian Telescope, Philips SPC900NC webcam, processed in Registax (Image by the author)

Jupiter

The same cannot be said of the moons of Jupiter. In theory four of them are bright enough to be seen by the naked eye [91]. Before the telescope was invented, nobody ever did, for the simple reason that Jupiter itself is so bright that our eyes cannot

Fig. 7.18 Mars, December 6, 2007. The visible features are similar to those of the previous night – a clue that the rotation period is about one Earth day. Taken using an 8-in. Newtonian Telescope, Philips SPC900NC webcam, processed in Registax (Image by the author)

Fig. 7.19 Mars, January 6, 2008. A different surface region is now facing Earth. Taken using an 8-in. Newtonian Telescope, Philips SPC900NC webcam, processed in Registax (Image by the author)

resolve them as separate objects. Jupiter whites out the neighborhood of its image on our retinas. It is the fourth brightest object in the sky after the Sun, the Moon and Venus.

You can see the four Galilean moons of Jupiter with binoculars. You do a lot better with tripod-mounted binoculars, but failing that, you can help by leaning

Fig. 7.20 Mars, February 11, 2008. By now, Mars is no longer in full phase and the south polar ice cap is pointing towards Earth. The Earth-Mars distance is increasing, with the planet's apparent size decreasing again. Taken using an 8-in. Newtonian Telescope, Philips SPC900NC webcam, processed in Registax (Image by the author)

against a wall or other suitably stiff and strong object. You can also see Jupiter's disk quite clearly through binoculars.

Through a telescope on a mount, with magnification of 25× or more, you can watch the moons move. You actually see a two-dimensional projection of three-dimensional motion onto the background sky. The moons appear to move north and south as well as east and west relative to the current right ascension and declination of Jupiter. In fact they are going around Jupiter's equator. We normally look at Jupiter from a slight angle, and so don't see the orbits of the moons edge-on. This is illustrated in Figs. 7.21 and 7.22.

The four Galilean moons are listed in Table 7.1, together with their orbital periods and distances from Jupiter. There are 60 other known moons, most of them discovered using spacecraft. Not even the fifth biggest is visible in amateur telescopes. It is an irregularly shaped lump of rock 50 miles across. Since the Great Red Spot on Jupiter is bigger than Earth, imagine how small a 50-mile-long boulder would look in comparison. The fifth moon's discoverer used a 36-in. telescope. Not many amateurs own one of those.

The innermost moon, Io, is about the same distance from the center of Jupiter as our Moon is from the center of Earth. There is so much more gravity pulling Io than the Moon that it takes less than 2 days, not a month, to complete one orbit. These moons are all fast moving. You can easily observe their movement over as little as 10 min, especially if you keep a photographic record. To do this, you have to over-expose the planet so that the moons can be seen. In theory you need at least a 10-in. telescope to resolve the Moons as discs. In practice you may need a larger aperture to allow for the limitations of the CCD chip. The pictures below were taken with

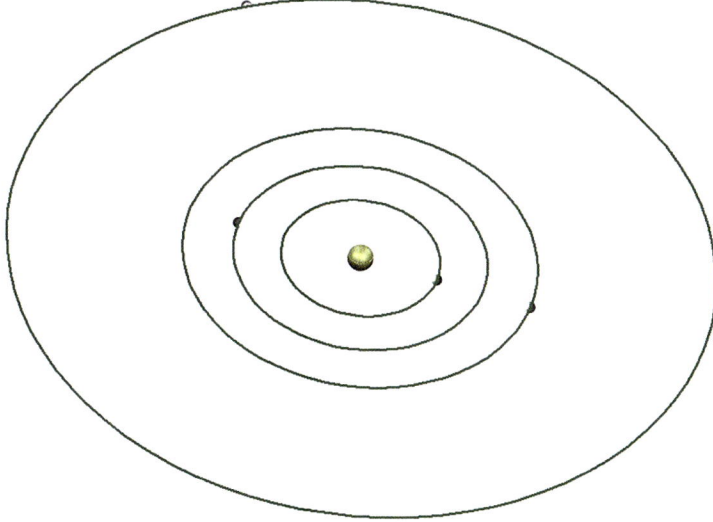

Fig. 7.21 The Jovian moons plying their orbits might be in the positions shown (Image by the author)

Fig. 7.22 This is where we might see the moons when looking from Earth. Note that they do not appear to be collinear, and they do not appear to align with the Jovian equator about which they orbit (Image by the author)

Table 7.1 Jupiter's moons

Moon	Orbital period (days)	Distance from Jupiter's center (thousands of miles)
Io	1.77	262
Europa	3.55	417
Ganymede	7.16	665
Callisto	16.69	1,170

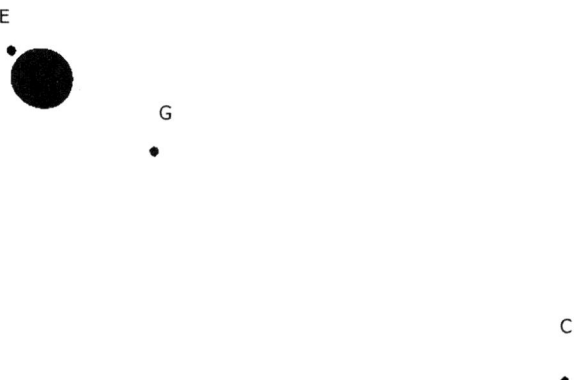

Fig. 7.23 Jupiter's moons, 2008-07-24 at 22:51 UT. Key: *I* Io (not visible), *E* Europa, *G* Ganymede, *C* Callisto. Taken using an 8-in. Newtonian telescope, Philips SPC900NC webcam, processed in Registax (Image by the author)

Fig. 7.24 Jupiter's moons, 2008-07-24 at 23:08 UT. Key: *I* Io (not visible), *E* Europa, *G* Ganymede, *C* Callisto. Taken using an 8-in. Newtonian Telescope, Philips SPC900NC webcam, processed in Registax (Image by the author)

an 8-in. telescope. I also cheated slightly and used the painting tool in *Photoshop*™ to allow for the inevitable loss of resolution in the book production process. Callisto is noticeably fainter than the other moons and would not otherwise have been visible. The images in Figs. 7.23, 7.24, 7.25, 7.26 and 7.27 are posterized negatives made from a color original in which the color was switched off.

A busy hour in July 2008 is illustrated in Figs. 7.23, 7.24, 7.25, 7.26 and 7.27. Europa has moved in front of Jupiter and become invisible. Io has emerged from behind Jupiter. Both Ganymede and Callisto have moved noticeably closer to Jupiter.

Fig. 7.25 Jupiter's moons, 2008-07-24 at 23:17 UT. Key: *I* Io (emerged from behind Jupiter), *E* Europa (disappearing in front of Jupiter), *G* Ganymede, *C* Callisto. Taken using an 8-in. Newtonian telescope, Philips SPC900NC webcam, processed in Registax (Image by the author)

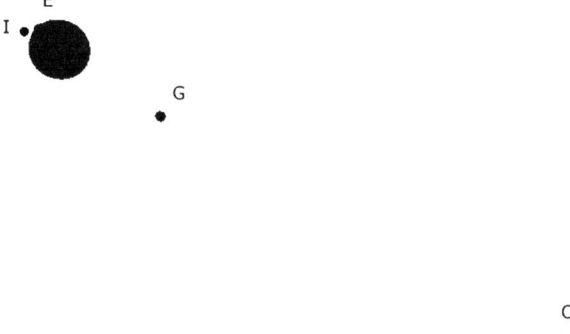

Fig. 7.26 Jupiter's moons, 2008-07-24 at 23:30 UT. Key: *I* Io, *E* Europa (nearly invisible), *G* Ganymede, *C* Callisto. Taken using an 8-in. Newtonian telescope, Philips SPC900NC webcam, processed in Registax (Image by the author)

Fig. 7.27 Jupiter's moons, 2008-07-24 at 23:43 UT. Key: *I* Io, *E* Europa (not visible), *G* Ganymede, *C* Callisto. Taken using an 8-in. Newtonian telescope, Philips SPC900NC webcam, processed in Registax (Image by the author)

Fig. 7.28 Jupiter's moons, 2009-07-14 at 01:55 UT. Key: *I* Io, *E* Europa, *G* Ganymede. Notice that Callisto is not visible. Unlike Figs. 7.23, 7.24, 7.25, 7.26 and 7.27, this photograph and the one that follows have not been posterized. They have been heavily processed in Registax and turned into negatives in Photoshop, but the differences in moon brightness are real. Taken using an 8-in. Newtonian telescope, Philips SPC900NC webcam (Image by the author)

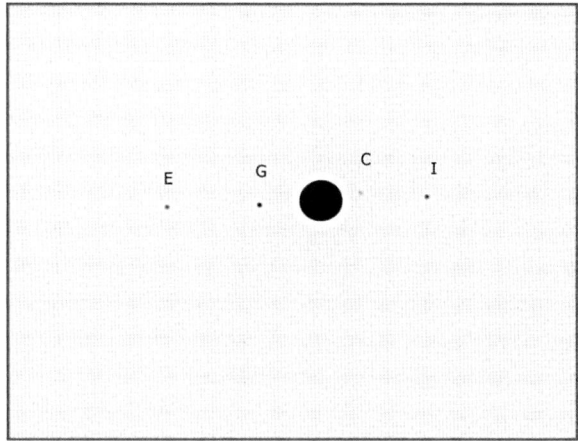

Fig. 7.29 Jupiter's moons, 2009-07-14, 14 min later at 02:09 UT. Key: *I* Io, *E* Europa, *G* Ganymede, *C* Callisto. Notice that Callisto has now appeared from the shadow cast by Jupiter. Taken using an 8-in. Newtonian telescope, Philips SPC900NC webcam, processed in Registax (Image by the author)

Figures 7.28 and 7.29, both negative images, show a rare event: the end of an eclipse of Callisto. Most years Jupiter's tilt relative to Earth means that Callisto is visible at all points in its 16⅔-day orbit. For a few of the 12 years of Jupiter's orbit this is not true.

Figure 7.29 also shows a way to identify the moons. It takes a little practice, but the order of brightness is: Ganymede, Io, Europa and Callisto. You can also cheat. There are lots of little apps around, such as this one from *Sky & Telescope*: http://www.skyandtelescope.com/observing/objects/javascript/jupiter#, which will tell you what to look for and which Moon is which.

What of the King of the Planets itself?

It is, of course, huge. It is so big, having well over twice the mass of the other planets put together [92], that some anonymous wit described the Solar System as consisting of the Sun, Jupiter and some other junk. It is a 'gas giant.' The sphere that we see is not solid. A ball of gas will not have a well-defined surface, like the surface of Mercury. What we actually see are ammonia clouds.

The mass of Jupiter [93] can be determined from the orbits of its moons. We know how far away it is, so we know how big it is. Hence we can determine its average density. It turns out to be about as dense as the Sun. Like the Sun it is made of 75 % hydrogen, 25 % helium. That, however, is where the similarity ends. The Sun is much hotter, and the hydrogen and helium exist as plasmas. In Jupiter, the pressure due to gravity is way beyond anything we experience on Earth. It squeezes the hydrogen into a state called 'metallic hydrogen,' [94] which has only ever been produced on Earth for very short times as a result of compression by high explosives [95]. There may or may not be a rocky core at the center – we just don't know [96].

There are tricks for measuring the mass at various distances from the center of the planet. These techniques rely on exploiting the oblateness of the planet, and work a lot better if you have a space probe you can use to fly around the planet, while you very accurately track its trajectory from Earth. From such models, people have tried to work out the mass of the rocky cores [97, 98]. It was found that Jupiter may have a smaller core than Saturn [99]. If the planets formed by gravitational collapse, this result is puzzling. It may have been the result of mergers of colliding planets or proto-planets. This is all fascinating, but alas as usual the theorists are really waiting for better data. The data we have don't rule out enough possibilities for us to eliminate some of them.

We obviously don't know much about metallic hydrogen because of the difficulty of making it on Earth. One or two detections in the whole of history, each at about a millionth of a second, do not exactly offer an opportunity for careful scientific investigation.

Jupiter, then, is an exotic planet by earthly standards. The part we do see, the atmospheric clouds, also displays out-of-this-world phenomena. The most famous of these is the Great Red Spot, a whirlwind stronger than any terrestrial hurricane, bigger than Earth, which has been blowing at least since we first had good enough telescopes to see it [100]. We do not know when it started. This spot is the feature of Jupiter that everyone wants to see. It is harder to see than people would have you believe, especially telescope sales people. First off, it is only really visible for about 1 h out of every 10. Jupiter rotates very fast. There are many internet apps that will tell you when it is around, such as http://www.skyandtelescope.com/observing/objects/javascript/3304091.html. The spot has shrunk over the last few decades, which does not help. It does not show up 'live' on my computer screen from a webcam on an 8-in. telescope, but it appears once processing with *Registax* starts. It can be seen by visual observation through a 12-in. telescope. Perhaps controversially, one might regard it as primarily a photographic object through smaller scopes.

In 2008 and 2009, Jupiter was very low in the English sky, never more than 16° above the horizon [101]. Good photography was just impossible. Figure 7.30 shows the first halfway decent picture obtained by the author, in 2010.

Figures 7.33 and 7.34 show the effect of photographing with a webcam at 5 frames per second and 15 frames per second, respectively. These pictures were

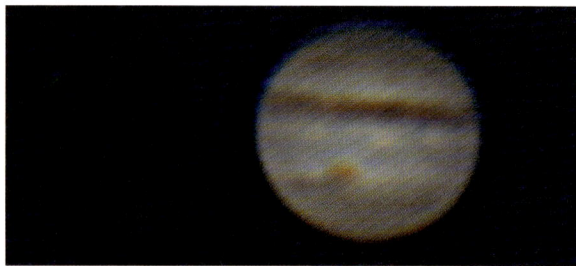

Fig. 7.30 Image of Jupiter, September 2, 2010, at 23:27 UT, through an 8-in. f/6 telescope, using a 3.4× Barlow lens. The moon to the *left* is Io. Taken using a Philips SPC900NC webcam, processed in Registax (Image by the author)

Fig. 7.31 View of Jupiter June 26, 2008, at 01:29 UT. Taken using an 8-in. Newtonian telescope, Philips SPC900NC webcam, processed in Registax (Image by the author)

taken to see what the trade-off was between data compression through the webcam's USB1 port and the effect of Jupiter's rapid rotation. In order to make the comparison as fair as possible, *Registax* was tuned to accept about half the frames in each case, and separate dark frames were shot for each picture. The seeing was rather poor, and the sky was slightly milky. A good test for night sky transparency in the Northern Hemisphere is whether you can you see iota Gemini. On clear nights you can see it in towns. On murky-ish nights you can't. It went in and out of sight that night. The result of the test is clear: despite the longer exposure time, 5 fps beats 15 fps with a webcam.

Notice also how far the planet has rotated in 11 min. That's why you have to be quick to catch the Great Red Spot.

Figures 7.30, 7.31, 7.32, 7.33 and 7.34 also show how Jupiter's clouds change over the years. The Great Red Spot is in Jupiter's southern hemisphere. The feature

Fig. 7.32 Image of Jupiter, Ganymede (*right*) and Callisto (*left*), September 19, 2011, at 00:11 UT. There is a lot of luck in getting a picture when the seeing is good. The Great Red Spot is barely visible. Taken using an 8-in. Newtonian telescope, Philips SPC900NC webcam, processed in Registax (Image by the author)

Fig. 7.33 Image of Jupiter shot at 15 fps. 450 frames were shot over 30 s. You cannot unambiguously tell whether the Great Red Spot is there, and there are spurious concentric lines on the picture. November 22, 2011, at 22:54 UT. Taken using an 8-in. Newtonian telescope, Philips SPC900NC webcam, processed in Registax (Image by the author)

just to the right (west) of it in the northern hemisphere in 2011 (Figs. 7.32, 7.33 and 7.34) was not there in earlier years. In 2008 and 2011 the darkest band in the southern hemisphere was to the north of the Great Red Spot (Figs. 7.31, 7.32, 7.33 and 7.34). In 2010 (Fig. 7.30) it was to the south and fainter. Part of the fun of observing Jupiter is that its clouds provide a rich source of variety.

Fig. 7.34 Image of Jupiter shot at 5 fps. 450 frames were over 90 s. Despite poor seeing, you can tell that the Great Red Spot is visible. There are no spurious concentric lines on the picture. November 22, 2011, at 23:05 UT. Taken using an 8-in. Newtonian telescope, Philips SPC900NC webcam, processed in Registax (Image by the author)

Saturn: The Iconic Planet

It is not so easy to see changes in Saturn's clouds. The planet is about twice as far away from the Sun as Jupiter, and therefore from us (Table 1.1), which makes it harder to see; and, the color changes are more subtle. It is cooler than Jupiter. Even though the chemical reactions giving rise to the colors are not fully understood, we can infer that the lower temperature means that Jupiter's clouds will be more active than those of Saturn.

What makes Saturn such a poster child for planets is of course those rings. We now know that all the gas giants have rings, but only those of Saturn were discovered without space probes. They were spotted by Christiaan Huygens, one of the great pioneers of modern physics, in 1655 [102]. Half a century after Galileo first looked at Saturn through a telescope, Huygens had a much superior instrument of his own design and could see that the planet was indeed surrounded by a ring. Huygens also showed how the tilt of the rings changes because the rotational axis of Saturn is inclined at 26.7° to its orbital plane. The rings orbit in the planet's equatorial plane. We therefore see them more or less tilted, over the 29.4 years it takes to complete an orbit. The last time they were edge-on was during the 2008–2009 apparition (Fig. 7.35).

The major influence on the apparent tilt of the rings as seen from Earth is the movement of Saturn along its orbital path (Fig. 7.36). Earth also moves along its orbital path, so the line of sight from Earth to Saturn is the result of the motion of the two planets.

Fig. 7.35 Saturn's rings orbit around the planet's equator. December 7, 2008, 07:02 UT. Taken using an 8-in. Newtonian telescope, Philips SPC900NC webcam, processed in Registax) (Image by the author)

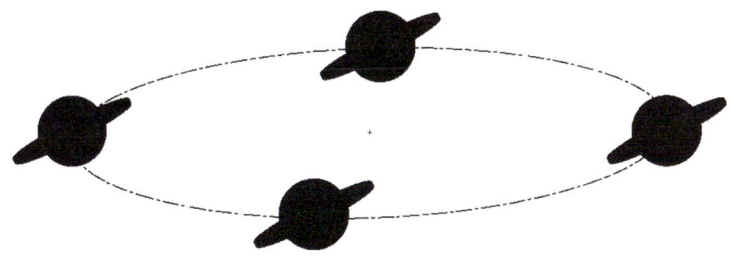

Fig. 7.36 The rings of Saturn remain parallel as the planet orbits. We see them appearing to tilt anywhere from 0° to about 29.3° (26.7° tilt of axis relative to Saturn's orbital plane, plus 2.5° inclination of orbital plane to ecliptic) (Image by the author)

Twenty years or so after Huygens, Cassini discovered that there are in fact several concentric rings [103]. It took almost another 200 years before James Clerk Maxwell, another physicist of the first rank, demonstrated that the rings cannot be solid or they would break up [104]. In effect he discovered that they consist of independent, orbiting particles, which we now know to be made mainly of ice.

Figures 7.35, 7.37, 7.38 and 7.39 show how the apparent tilt of Saturn's rings changes over the years.

It is also worth noting that the quality of these photos is very variable. That happens a lot. Figure 7.39 is included with some reluctance because it is so poor. The images obtained are not usually as bad as Fig. 7.39, but to get a good quality image such as shown in Fig. 7.38 is rare. The seeing is not usually that good.

Fig. 7.37 Note the tilt of Saturn's rings in 2007 (March 9). Taken using a 6-in. f/5 Newtonian telescope, Philips SPC900NC webcam, processed in Registax (Image by the author)

Fig. 7.38 Note the tilt of Saturn's rings in 2008. February 9, 2008, 00:38 UT. Taken using an 8-in. Newtonian telescope, Philips SPC900NC webcam, processed in Registax (Image by the author)

It is important that you realize that you are not often going to get pictures as good as the ones published in the monthly magazines. Otherwise you will become demoralized and think that you are no good. Unfortunately the weakest link in our optics is the atmosphere.

Saturn has quite a family of moons. Titan is easily visible with a 6-in. reflector by eye. It really needs at least an 8-in. instrument to see the other moons.

Fig. 7.39 Note the tilt of Saturn's rings in 2011. April 30, 2011, 22:51 UT. Unlike in Figs. 7.37 and 7.38, we are looking at the rings from the north, not the south. Compared to Fig. 7.38, the quality of the image is poor. That's the difference between good and bad seeing. Taken using an 8-in. Newtonian telescope, Philips SPC900NC webcam, processed in Registax (Image by the author)

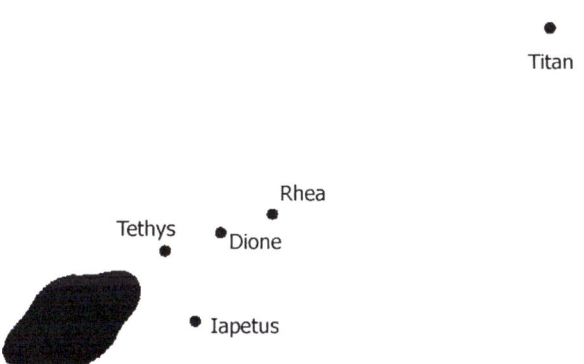

Fig. 7.40 Saturn and some of its moons. February 10, 2008, 21:34 UT. Picture taken with 8-in. f/6 Newtonian, with a 2.3× Barlow lens and a Philips SPC900NC webcam (Image by the author)

Figures 7.40 and 7.43 are from photos taken on the same night. Over the course of February 10–11, 2008, the moons Dione and Tethys passed one another. Rhea, which has a period of 4.5 days, moved appreciably towards Saturn. Of course it is in a nearly circular orbit, so it only appears to approach Saturn. Titan and Iapetus did not appear to move much. Titan has a period of 16 days and Iapetus of 79 days.

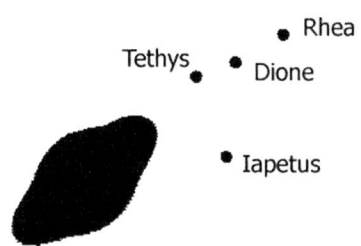

Titan

Rhea
Tethys   Dione

Iapetus

Fig. 7.41 Saturn and some of its moons. February 10, 2008, 23:30 UT. Picture taken with 8-in. f/6 Newtonian, with a 2.3× Barlow lens and a Philips SPC900NC webcam (Image by the author)

Iapetus does not orbit in the plane of the rings, but the other moons do. Iapetus also has the odd characteristic that it has one bright side and one darker side. When we see it traveling from east to west (left to right in the image) it is bright. A webcam in an 8-in. telescope does not see it when it makes the return journey. Iapetus is not then bright enough.

Indeed you are operating at the limit of your capability photographing Saturn's moons. Depending on the atmospheric transparency, the brightness and often the gamma and contrast controls have to be turned most or even all the way up. To follow Titan's orbit, it is necessary to switch between no Barlow lens, a 2.3× Barlow and a 3.4× Barlow lens as Titan appears to move closer to and further away from Saturn.

Unusually for a gas giant's moon, most of which appear white in photos, Titan shows a hint of orange too often for it to be coincidence.

Titan looks orange because it has an atmosphere of mostly methane. It is the only moon in the Solar System to possess a significant atmosphere [105]. It is also the only moon of a gas giant to have been visited by a soft-landing probe, the European Space Agency's Huygens probe, which traveled to Saturn with the incredibly successful Cassini orbiter [106]. Indeed the Cassini-Huygens probe confirmed the composition of the atmosphere. Its pressure and density are higher than those at Earth's surface [107]. With distressing predictability, there was much press hullaballoo about life on Titan at the time of the Huygens landing. The temperatures measured by the Huygens probe were about −150 °C [108]. Life cannot exist at that temperature. It's over 130 °C cooler than a deep freeze (Figs. 7.41, 7.42 and 7.43).

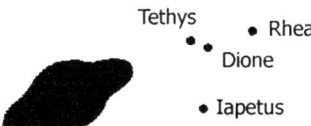

Fig. 7.42 Saturn and some of its moons. February 11, 2008, 01:38 UT. Picture taken with 8-in. f/6 Newtonian, with a 2.3× Barlow lens and a Philips SPC900NC webcam. Registax (Image by the author)

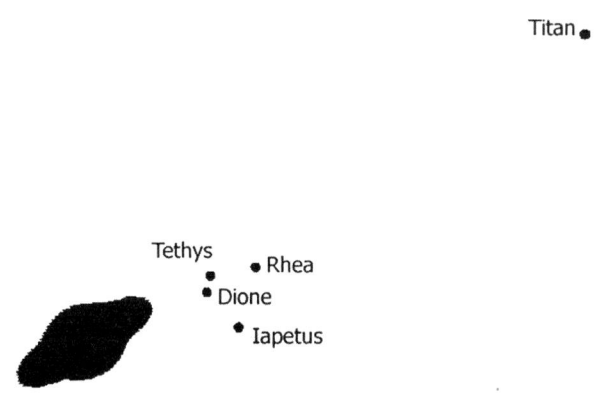

Fig. 7.43 Saturn and some of its moons. February 11, 2008, 03:47 UT. Picture taken with 8-in. f/6 Newtonian, with a 2.3× Barlow lens and a Philips SPC900NC webcam (Image by the author)

Do the moons of Saturn, apart from Iapetus, orbit in the plane of the rings, the equatorial plane of Saturn? In the winter of 2007–2008 this was not altogether obvious, as Fig. 7.44 suggests.

The chance to find out came in December 2008, when the rings were edge-on. Figures 7.45, 7.46, 7.47, 7.48 and 7.49 show the results of following the moons through approximately one complete orbit of Titan. The moons shown do indeed appear to orbit in the plane of the rings.

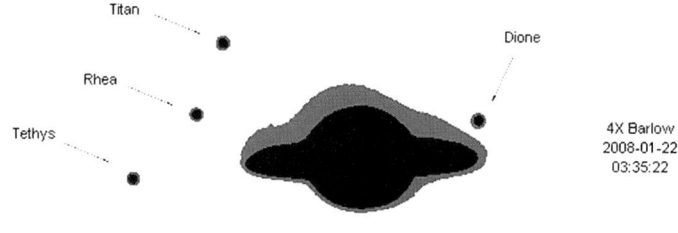

Satellite Movement over 50 minutes

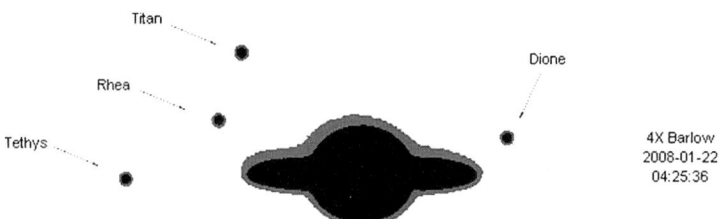

Fig. 7.44 Saturn and some of its moons, showing how it is not obvious where the orbital planes are. January 22, 2008, times shown are UT. Picture taken with 8-in. f/6 Newtonian, with a Philips SPC900NC webcam. Processed in Registax. Satellites enhanced using Photoshop. The '4×' Barlow turned out to be 3.4× (Image by the author)

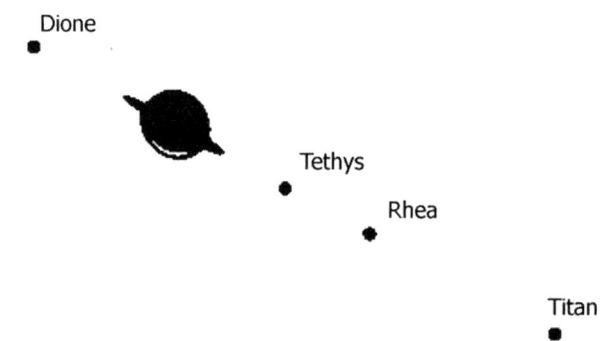

Fig. 7.45 Saturn and some of its moons, looking edge-on. December 9, 2008, 06:09 UT. Picture taken with 8-in. f/6 Newtonian, with a 2.3× Barlow lens and a Philips SPC900NC webcam. Processed in Registax. Satellites enhanced using Photoshop (Image by the author)

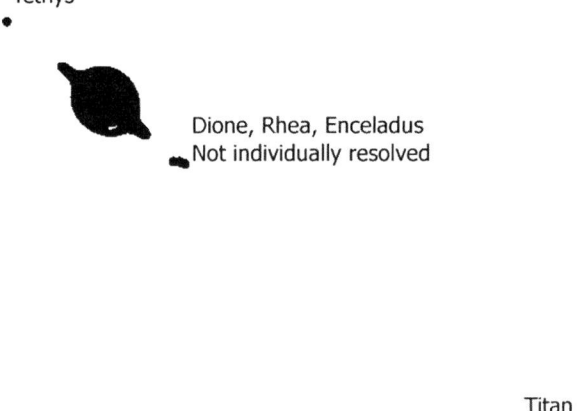

Fig. 7.46 Saturn and some of its moons, looking edge-on. December 10, 2008, 06:36 UT. Picture taken with 8-in. f/6 Newtonian, with a 2.3× Barlow lens and a Philips SPC900NC webcam. Processed in Registax. Satellites enhanced using Photoshop (Image by the author)

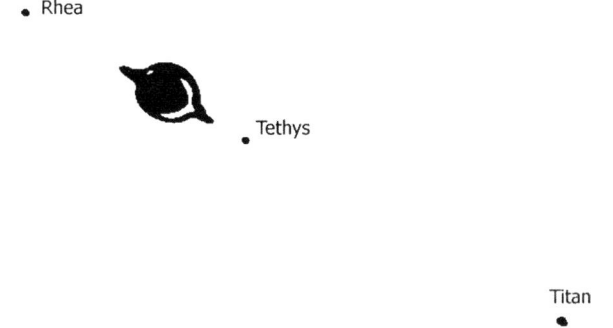

Fig. 7.47 Saturn and some of its moons, looking edge-on. December 11, 2008, 06:57 UT. Picture taken with 8-in. f/6 Newtonian, with a 2.3× Barlow lens and a Philips SPC900NC webcam. Processed in Registax. Satellites enhanced using Photoshop (Image by the author)

The moons other than Titan and occasionally Rhea appear very similar, so much so that it really requires an Internet app to identify the moons: http://www.skyandtelescope.com/observing/objects/javascript/saturn_moons#.

Finally, you will notice some white arcs in photos taken with the Barlow lens, as the one used was of poor quality. Its achromatic abilities were not impressive, and this leads to all sorts of fringing.

Fig. 7.48 Saturn and some of its moons, looking edge-on. December 17, 2008, 04:11 UT. Picture taken with 8-in. f/6 Newtonian, with a 2.3× Barlow lens and a Philips SPC900NC webcam. Processed in Registax. Satellites enhanced using Photoshop (Image by the author)

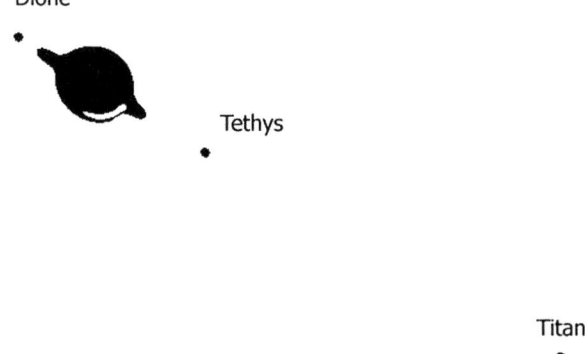

Fig. 7.49 Saturn and some of its moons, looking edge-on. December 26, 2008, 05:35 UT. Picture taken with 8-in. f/6 Newtonian, with a 2.3× Barlow lens and a Philips SPC900NC webcam. Processed in Registax. Satellites enhanced using Photoshop (Image by the author)

The Telescope Planets: Uranus and Neptune

It is often claimed that Uranus and Neptune are binocular objects, and that Uranus can even be a naked-eye object. What they rarely bother to add is that they are binocular objects if you know what and where they are. They are not really distinguishable from stars in binoculars, especially in a street-lit area. With an 8-in. telescope you can tell that they are discs; and they have that planet-like quality of taking longer to disappear into the dawn sky than stars do. The author first found these two planets with a GOTO system. The main method of confirming that they

Fig. 7.50 A conjunction of Venus and Uranus. I happened upon this by accident. I had been tracking the orbit of Uranus all that autumn and knew what the asterisms around it looked like. I unexpectedly recognized them in what I thought would be a photo of Venus. Taken using a Canon EOS 100D DSLR, 200 mm lens, f/5.6 (Image by the author)

are planets is to observe them move over a few nights. Stars don't do that to any appreciable extent. Figures 7.50 and 7.51 show planetary motion. Lurking in these photos of Venus is another moving object – Uranus.

A webcam image of Uranus, Fig. 7.52, was disappointing. You can see that it is a disc, but that is about all. This poor picture fits the chapter's theme: what you can realistically expect to see. If you can do better, great. But this shows why you see beautiful pictures of Jupiter and Saturn in the magazines, but not often of Uranus or Neptune.

In 2009 Neptune had a triple conjunction with Jupiter. Conjunctions are not that rare, because the planetary orbits are almost co-planar. If the planes of the orbits were random, conjunctions would be extremely rare.

Figures 7.53, 7.54 and 7.55 show both Jupiter and Neptune against the starry background. You can see that Neptune moves slightly but noticeably in 1 day and moves very noticeably over four days. Figure 7.55 is less clear; the atmospheric conditions were less favorable that night. The waning crescent Moon had yet to rise, so scattered moonlight was not the cause. If the Moon is making a nuisance of itself, it illuminates the slightest hint of cloud, and you see a lot fewer stars.

As has been mentioned elsewhere, you will only find planets close to an imaginary line in the sky called the ecliptic. If you are adventurous and willing to do a little geometry, you can work out the orbits of the planets for yourself. Some very simple methods are given in the companion book in this series by Clark [109].

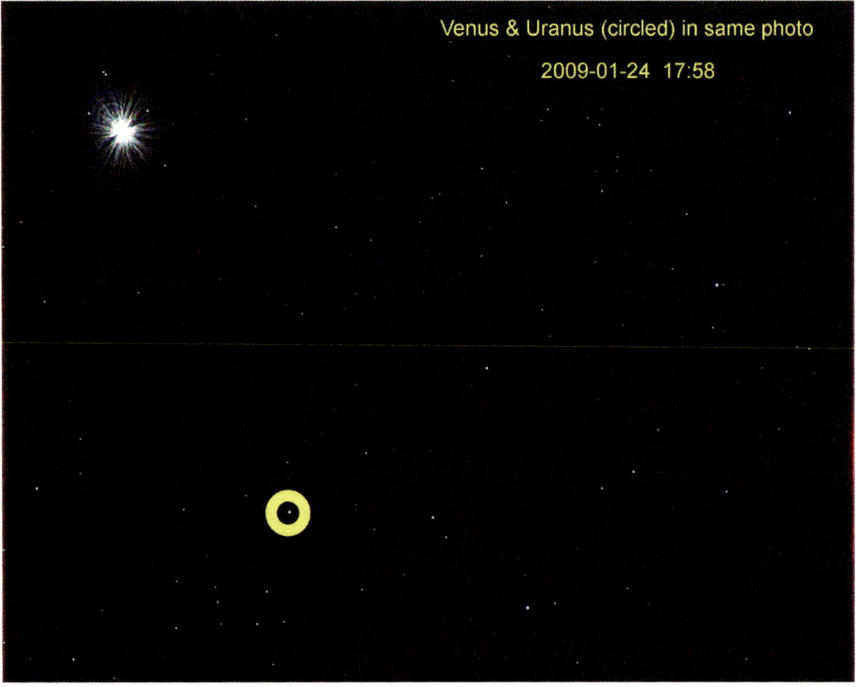

Venus & Uranus (circled) in same photo

2009-01-24 17:58

Fig. 7.51 A few days after taking the picture shown in Fig. 7.50, I took another one to test my suspicion that I had found Uranus. The confirmation is that it had moved to the east in the intervening 4 days. Canon EOS 100D DSLR, 200 mm lens, f/5.6 (Image by the author)

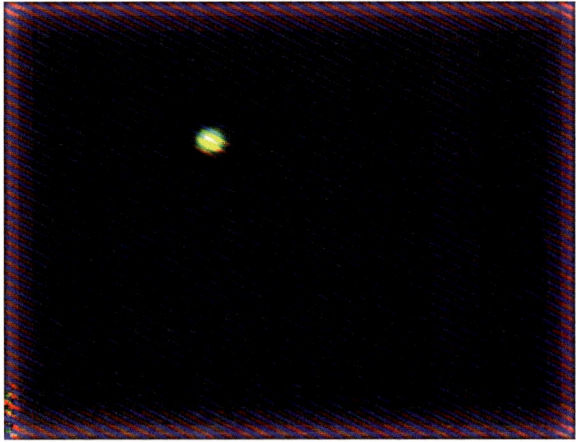

Fig. 7.52 A webcam photo of Uranus. You do not see much. You could estimate its angular diameter from this image, but I am not inclined to trust the result. October 26, 2008, 21:42 UT. Picture taken with 8-in. f/6 Newtonian, with a 3.4× Barlow lens and a Philips SPC900NC webcam. Processed in Registax (Image by the author)

Fig. 7.53 A DSLR photo of Neptune and Jupiter, taken in the early hours of July 15, 2009. The DSLR was piggybacked onto a 6-in. Newtonian, into which a webcam had been inserted for guiding. For obvious reasons the object guided on was Jupiter. 500 mm lens, f/8. Exposure time: 5 min. The picture is the JPEG produced by the camera (Image by the author)

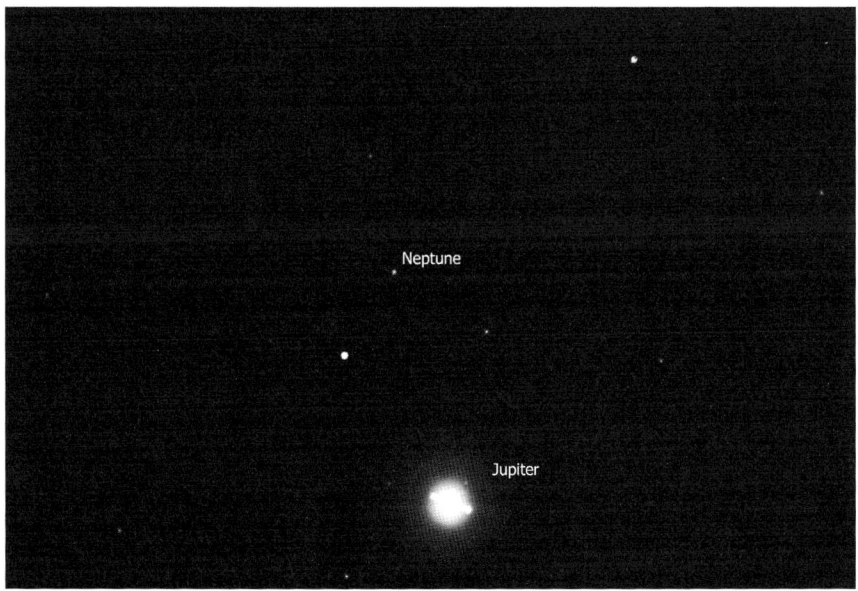

Fig. 7.54 A DSLR photo of Neptune and Jupiter, taken in the early hours of July 16, 2009. The DSLR was piggybacked onto a 6-in. Newtonian, into which a webcam had been inserted for guiding. For obvious reasons the object guided on was Jupiter. 500 mm lens, f/8. Exposure time: 5 min. The picture is the JPEG produced by the camera (Image by the author)

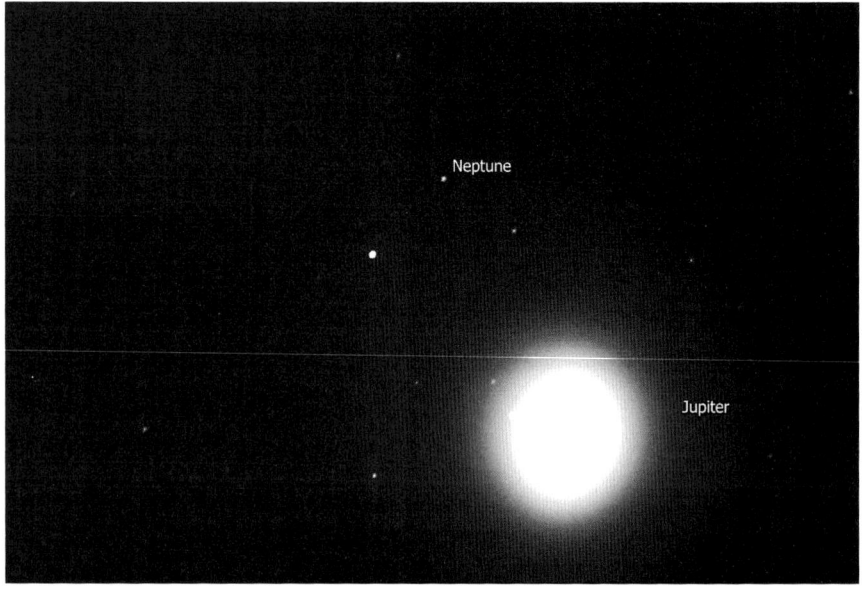

Fig. 7.55 A DSLR photo of Neptune and Jupiter, taken in the early hours of July 19 2009. The DSLR was piggybacked onto a 6″ Newtonian, into which a webcam had been inserted a webcam for guiding. For obvious reasons the object guided on was Jupiter. 500 mm leans, f/8. Exposure time: 5 min. The picture is the JPEG produced by the camera (Image by the author)

Chapter 8

The Sun, Star of the Solar System

The physics of the Sun is fairly well understood [110]. We would understand it even better if we had a decent theory of convection, which turns out to be the main energy transport mechanism between the nuclear reactions at the center and the outside world [111]. The chapter shows what you can expect to record for yourself; and shows how such tools as spectroscopy and helioseismology tech us much of what we know.

When discussing the solar eclipse of 1999, it was shown in Fig. 6.41 how you can project images of the Sun to render telescopic viewing safe. There are other techniques. You can place a film of filter material over the front of your telescope (Fig. 8.3). This film is not expensive to buy if you do not mind making a holder for it. It goes without saying that if you do this, inspect the film every time you use it for pinholes, and either repair them with opaque tape or replace the film.

There is a joint European Space Agency/NASA solar observatory called 'SOHO' in orbit at one of the 'Lagrange points' of Earth's orbit, a point moving with Earth where small objects are gravitationally stable. The SOHO website http://sohowww.nascom.nasa.gov shows up-to-date pictures.

The kinds of things you might see for yourself are shown in Fig. 8.1. The black regions are sunspots, which are only relatively dark. They are slightly cooler than the surrounding regions. The white regions are called 'plages.' The Sun's surface is grainy. The grains are convection cells. If you boil a pot of water, you can see that there are about four separate convection cells, regions of upwelling and downwelling, as the heating proceeds. You can often see many more granules than are visible in Fig. 8.1. The sunspots are often the size of Earth or thereabouts.

© Springer Science+Business Media New York 2015
J. Clark, *Viewing and Imaging the Solar System*, The Patrick Moore
Practical Astronomy Series, DOI 10.1007/978-1-4614-5179-2_8

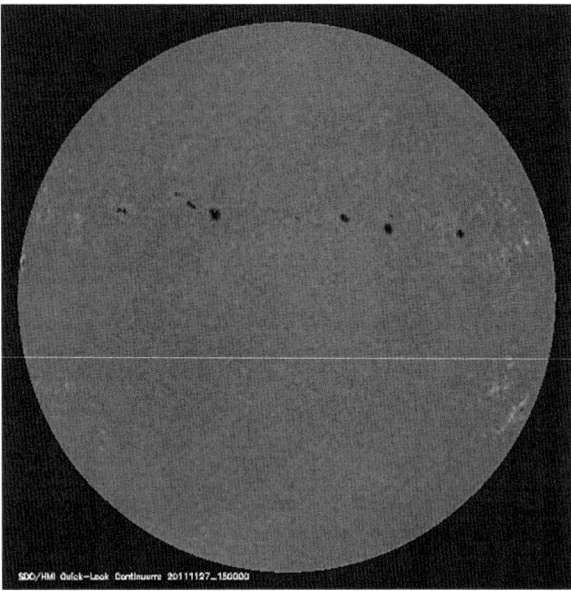

SDO/HMI Quick-Look Continuum 20111127_150000

Fig. 8.1 SOHO image at a monochrome wavelength called Ni I. This shows sunspots very well. The white features shown are called 'plages.' November 27, 2011, 15:00 UT (Image courtesy of ESA/NASA)

Figure 8.2 shows a SOHO image of the Sun in the far ultraviolet, taken at about the same time as Fig. 8.1. What a different animal the Sun then looks! The bright spots clearly correspond to the sunspots and plages.

You can become an armchair observer of the Sun nowadays and become an avid follower of the SOHO images (see Chap. 11). There is much to be said for this. They are never obscured by clouds, and you see much more than you ever could by eye.

However, it is more fun to see for yourself, the author claims, in spite of remembering with a shudder a recent foray into solar photography. The pictures were taken with an 8-in. telescope, with suitable protecting filters, using a webcam in afocal projection. All this is shown in Fig. 8.3.

The first challenge is to find the Sun without blinding yourself. You do this by first minimizing the shadow cast by the optical tube assembly, then with the filter and eyepiece removed, projecting what came out of the focuser onto your hand. Once the brightness of the mark on your hand is maximized, replace the filter and use an eyepiece with cross-hairs to fine-tune the alignment. Then insert the eyepiece with attached webcam. If you suffer from migraines, you will know that that kind of thing is a very good way to trigger one. Two little challenges ensue. The first is that at 51 °N the November Sun is low enough in the sky that trees get in the way. Challenge number two is how to focus the webcam image when the Sun is

2011/11/27 13:13

Fig. 8.2 SOHO image at far-UV wavelength: 195 Å. This shows sunspots and plages very well. November 27 2011 13:13 UT (Image courtesy of ESA/NASA)

Fig. 8.3 Telescope with film in place and webcam in afocal projection, behind an eyepiece (Image by the author)

shining onto your laptop screen, and about all you can see is the dust on it. It is better, if you have the luxury, to pre-focus on a bright star while it is still dark.

If you have to focus in daylight, one trick is to throw a towel over your head and over the screen on which you are trying to focus an image, to block out some of the sunlight. A dark towel would obviously be more opaque than a pale-colored one.

Fig. 8.4 The Sun, November 27, 2011, 14:22 UT. This image was taken at about the same time as Figs. 8.1 and 8.2. My orientation is different. I could have cured that by rotating my webcam in the focuser if I had wanted to. Nevertheless, you can see that I did capture most of the sunspots, but cannot claim to have photographed any plages or granules (Image by the author)

The result is shown in Fig. 8.4. The recipe was: shoot 30 s' worth of avi footage at 5 fps, and 30 s of dark frame, and combine them in *Registax*. Figures 8.1 and 8.2 were chosen to be as simultaneous with Fig. 8.4 as possible, to enable comparison.

Those sunspots not near the edge of the disk were captured, but little else. The purpose of showing you this less-than-wonderful image is to pitch your expectations of what you are likely to observe for yourself.

Any telescope can be used to project an image of the Sun onto paper. If the telescope is indoors, you can project onto a house wall. All you need to do is to have the eyepiece further back than you would to look through it, and it will project a real image. Be careful to have an eyepiece in the telescope. Projecting very concentrated sunlight from the objective mirror or lens of a telescope is a good way to start a fire!

Better results can be obtained with a DSLR camera than with a webcam. To enable a fair comparison, Figs. 8.5 and 8.6 were taken using different cameras but the same telescope, viz. a 102 mm f/10 Maksutov. The DSLR image in Fig. 8.6 is by far the better. The principal challenge in such DSLR photography is focusing. One possibility is to pre-focus the night before on a bright star using a Bahtinov mask. Another is to focus on a distant horizon object and use this as a basis, taking several pictures as the focus is slightly adjusted, then rejecting all but the best image (Figs. 8.7 and 8.8).

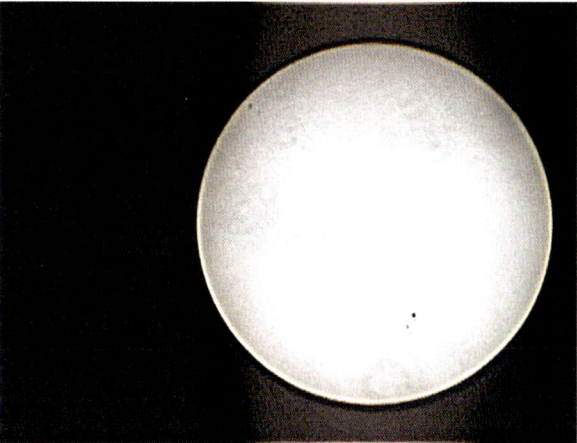

Fig. 8.5 The Sun, May 26, 2012, 07:44 UT. This image was taken with a webcam in afocal projection behind the eyepiece and a 102 mm (4 in.) f/10 Maksutov telescope on a computer-controlled alt-az mount, using a solar *white* light filter (Image by the author)

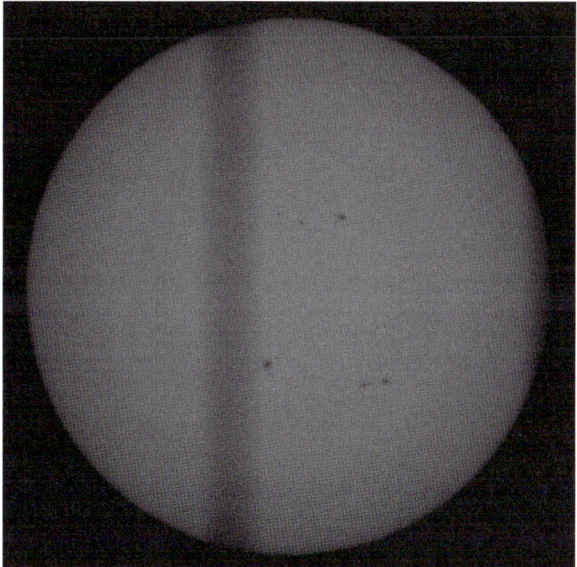

Fig. 8.6 The Sun, June 6, 2012, 04:46 UT. This image was taken with a DSLR in prime focus using the same 102-mm (4 in.) f/10 Maksutov telescope as Fig. 8.5. Exposure time: 1/3200 s (Image by the author)

Fig. 8.7 The arrangement used to photograph Fig. 8.6, with a DSLR in prime focus (Image by the author)

The Sun is so bright at magnitude −27 that you do not need a big telescope to see it. Indeed there is a case against using large telescopes: the risk of frying either your eyes or your equipment becomes that much greater. If you only have a large telescope, it is advisable to block off most of the objective area. Some telescope caps have a facility to do this (Fig. 8.9). You MUST still use a solar observing filter, even with such a cap.

There is a small telescope called a Coronado PST, which stands for 'Personal Solar Telescope.' This telescope is good for observing. You can easily see solar prominences with it. At the time of writing, this telescope retails for about $500 in the United States (Fig. 8.10).

This telescope has a very narrowband filter in it that only looks at the hydrogen alpha wavelength, as well as a safety filter. At this wavelength, you can easily see sunspots on the Sun's disc, and solar flares around the edge of the photosphere. You should be able to see plages. It costs a lot more money to get a good view of granulation.

The Coronado PST is the only inexpensive way you can see solar flares. Hydrogen alpha filters for your own telescope cost a lot more. Where the Coronado PST falls down is when it is used for photography. The real issue is that it lacks back-focus. You cannot get a camera far enough back to get an image of the whole star, even in afocal projection. There are some ingenious camera modifications on the Internet, such as the one at http://coronado-pst.blogspot.com, for a webcam at prime focus.

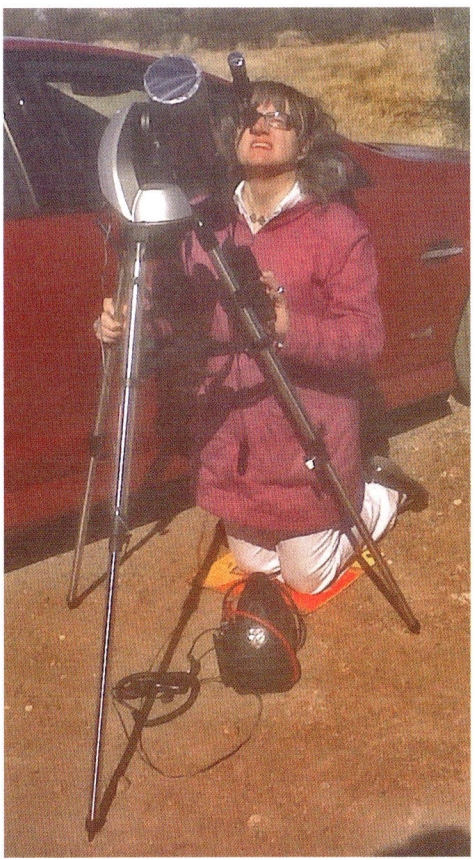

Fig. 8.8 Telescope with film in place, with a DSLR in prime focus (Image courtesy of Melissa Hulbert, used with permission)

Figure 8.11 was taken with a Coronado PST and a DSLR camera. Focus was achieved by throwing a towel over the photographer's head and the camera. Otherwise nothing could be seen for reflections off the photographer's face. Several exposures were taken over different times. Even so, focusing was far from easy. The precise exposure will depend on how high the Sun is in the sky and on atmospheric clarity. The most effective strategy is to try several exposures in quick succession and keep only the best image.

The surface 'roughness' seen in Fig. 8.11 is spurious. The same pattern could be seen on photographs taken the following day. It is presumably caused by the optics in the telescope.

A more expensive way to achieve such photography is with a Lunt solar telescope (Fig. 8.12). This telescope also photographs at the hydrogen alpha

Fig. 8.9 A telescope lens cap with a hole in it to make solar viewing safer. You MUST still cover this hole with a solar viewing filter (Image by the author)

Fig. 8.10 The Bristol Astronomical Society's Coronado PST solar viewing telescope on the mount of a member (Image by the author)

wavelength. Exposure times required are of the order of one second. The non-tracking mount shown is adequate for such exposures. A shot that shows the flares around the edge of the Sun's disk requires more exposure than a shot that would best show the surface granulation.

Fig. 8.11 A photo taken with the Coronado PST solar telescope with a DSLR, 31 May 2013, 14:11 UT, in afocal projection, 1/60 s exposure, HEQ5 mount (Image by the author)

Not only that but you also need to tune the wavelength passed by the filter slightly to see the corona or surface features. In fact both the Coronado and Lunt instruments use tunable solar filters described in the promotional literature by their French name of *etalon*. Stripped of the mystique, this is a Fabry-Perot interferometer, in which the wavelike properties of the light are used to cause unwanted wavelengths to be canceled out by interference.

The Sunspot Cycle

Sunspots were first observed telescopically by Galileo and his contemporaries. It was subsequently discovered [115] that the number of sunspots seen varies in a roughly 11-year cycle. Heinrich Schwabe, who noticed this, was actually looking for the postulated planet Vulcan. Of course he did not find it because it was not there, but he did see a lot of sunspots while he was waiting.

Figure 8.13 shows that the number of visible sunspots cycles over time. It is perhaps fortunate that sunspots were discovered before the onset of the Maunder minimum, when there were virtually no sunspots; otherwise we might think that they had appeared out of nowhere in the mid-1700s. It is correspondingly unfortunate that we do not know in detail what happened before the Maunder minimum.

Figure 8.14 shows other ways of looking at the sunspot cycle. Instead of counting the number of sunspots, the lower diagram shows the total area of the visible

Fig. 8.12 A Lunt solar telescope with a DSLR (Image by the author)

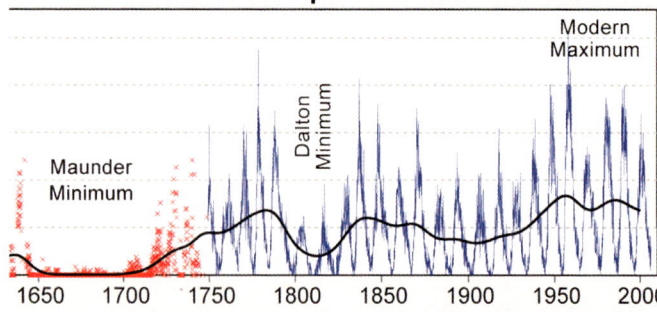

Fig. 8.13 The number of sunspots observed over the years. It can be seen that this number cycles about every 11 years, and there are longer-range variations, which do not seem to show much of a pattern (Image by Robert A. Rhode, Creative Commons Attribution-Share Alike 3.0 Unported license. http://en.wikipedia.org/wiki/File:Sunspot_Numbers.png)

NSPOT AREA AVERAGED OVER INDIVIDUAL SOLAR RC

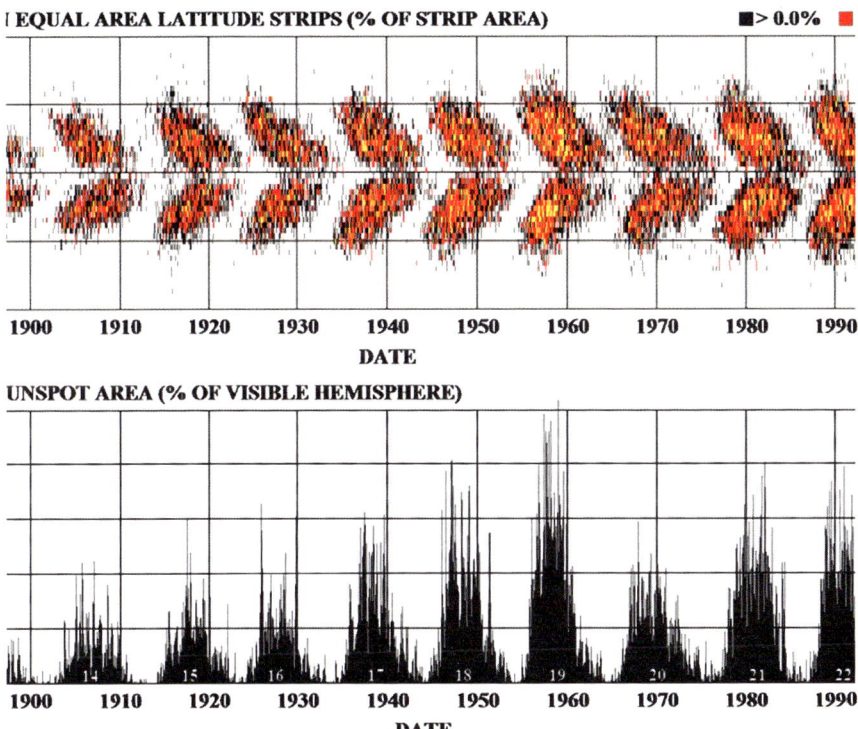

Fig. 8.14 The sunspot cycle in more detail. *Top*: The variation in latitude of sunspots over time. *Bottom*: The area of the visible Sun's disk covered with sunspots over time (Image courtesy of NASA)

solar disk that was covered by sunspots. The upper diagram shows yet another remarkable phenomenon – the latitude at which sunspots are observed declines as a solar cycle progresses.

If you are patient, you can verify these observations for yourself over the years.

Solar Spectra

What is interesting about solar spectra? By answering the question, "How bright is sunlight at each wavelength?" we can deduce how hot the solar photosphere is and what its chemical composition is.

Isaac Newton [116] was by no means the first to use a glass prism to break sunlight into its spectrum, or range of colors, or range of wavelengths since each

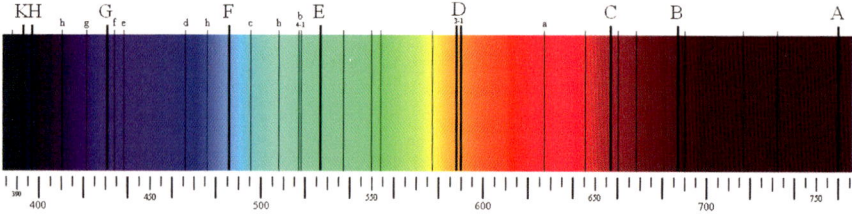

Fig. 8.15 The spectrum of the Sun. As expected, it contains the colors of the rainbow. Less intuitively obviously, it also contains some missing colors, which appear as *dark lines* in the spectrum. This diagram represents the range that most people can see well [117]. The unit of wavelength, the nanometer, or nm, is a billionth of a meter. There are 25.4 million nanometers in an inch. This is indeed a small unit (Image courtesy of http://en.wikipedia.org/wiki/Fraunhofer_lines#mediaviewer/File:Fraunhofer_lines.svg)

wavelength of light causes one color in the rainbow, but he certainly documented the phenomenon more thoroughly than his predecessors. Above all, he was the first to understand that white light is composed of primary colors, and is not a primary color in its own right. In other words, he was the first to realize that the colors of the spectrum were present in the light and could be separated.

Diffraction gratings, or sets of very closely, but evenly spaced, slits, will produce a spectrum. The theory of wave optics shows how such a grating also makes it possible to measure the wavelength of the light. [118] The first person to produce a diffraction grating, using hairs, seems to have been Benjamin Franklin's friend David Rittenhouse [119], who was the first director of the U. S. Mint as well as an inventor and astronomer. (Curiously, Isaac Newton also became Master of the British Mint.) Rittenhouse estimated from his diffraction grating the wavelength of light. Scotsman Thomas Young quantified the relationship between slit spacing and angle through which a given wavelength of light is diffracted [120]. The German physicist Joseph von Fraunhofer built a similar grating in 1821 and discovered the dark lines shown in Fig. 8.15 [121].

Diffraction gratings work as follows. If you shine light through two slits, each of the same size order as the wavelength of the light, you will see a phenomenon called interference. The principle of interference is simple enough. If two waves pass through one another, their amplitudes at any point add. If they are in phase, you get constructive interference, as shown in Fig. 8.16. If they are exactly out of phase (in *antiphase*), they cancel, giving destructive interference, shown in Fig. 8.17.

If a wave passes through a wall with gaps, or slits, in it; and if these gaps are of the same order of size as the wavelength, then the wave emerges from each gap as if it had originated there, as shown in Fig. 8.18. This phenomenon is called Huygens' principle [122]. It is named after the same Huygens who discovered Saturn's moon Titan.

Where the two emergent waves are in phase, constructive interference occurs. Where they are in antiphase, the waves cancel.

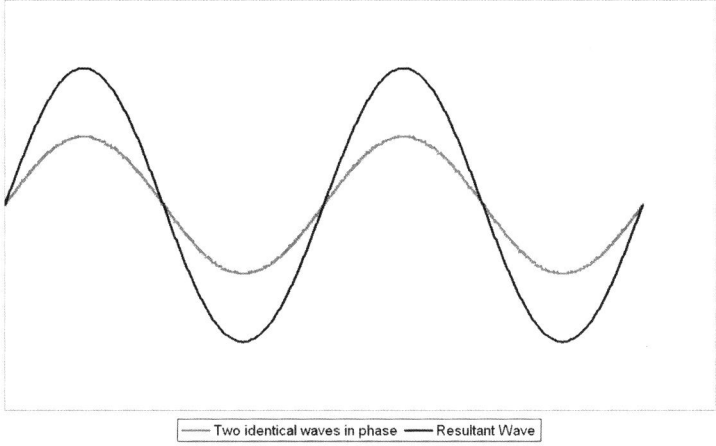

Fig. 8.16 Two identical waves in phase with one another add up to the wave shown as the *darker line*, which is what an observer would detect. This is called constructive interference (Image by the author)

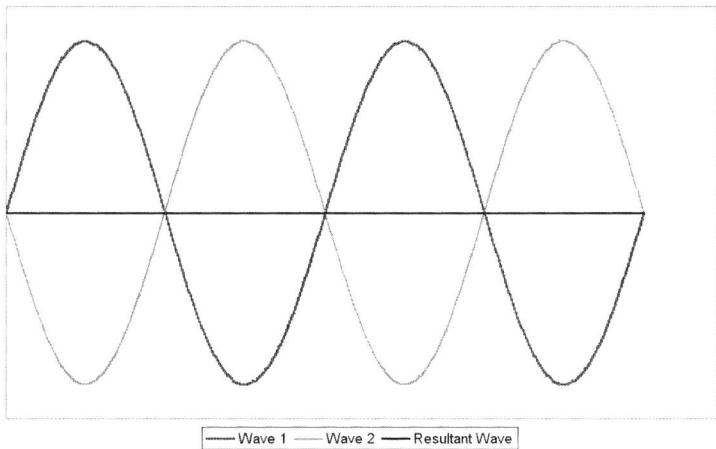

Fig. 8.17 Two identical waves with opposite phases (in "antiphase") add up to zero net wave, which is what an observer would detect. This is called destructive interference (Image by the author)

A diffraction grating is an arrangement like that in Fig. 8.18, except that the number of slits is very large. If they are identically spaced, then there is only a net wave of a given wavelength in a handful of directions. The longer the wavelength, the greater the angle around which the wave diffracts. This is why white light splits into its constituent spectrum when passed through a diffraction grating.

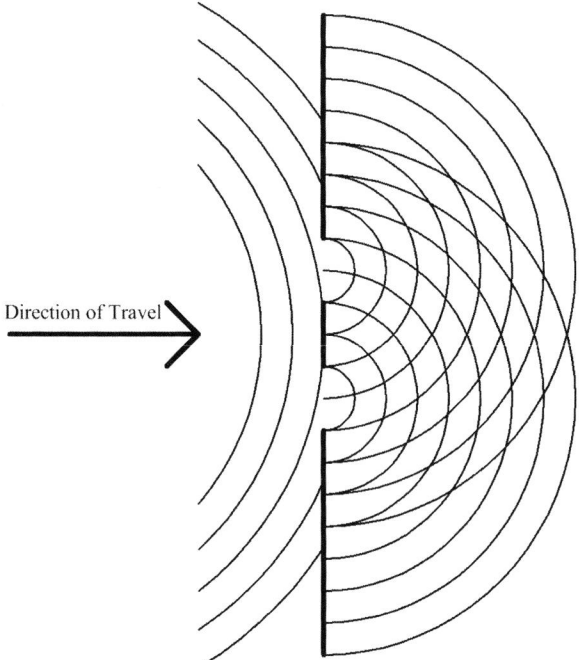

Fig. 8.18 When a wave traveling from left to right reaches a wall with two slits, the downstream wave from each slit appears as if it were a wave originating at the slit. This is known as Huygens' principle. There would also be waves reflected upstream of the wall. These have been omitted for clarity (Image by the author)

You can get transmission diffraction gratings and reflective diffraction gratings (Fig. 8.19).

We will show how these curious dark lines appear by considering the simplest case of a hydrogen atom. Hydrogen atoms only have one electron, which eliminates the additional complication of mutual repulsion between negatively charged electrons. Nevertheless, the principle is the same for more complex atoms. It is just that the mathematics become far harder – not that we will get into mathematics.

The temperature of the surface of a star can be determined by measuring the relative abundance of light at various wavelengths and comparing this to Planck's law of black body radiation [123]. The amount of radiation present at different wavelengths at different temperatures is shown in Fig. 8.20. The thickest line, at 5,500 °C, represents the approximate temperature of the photosphere of the Sun, the visible surface. The vertical lines delineate the visible range. Note that the wavelength ranges shown exceed the visible range.

We next discuss the origin of the dark lines in the spectrum of the Sun. Then we will discuss what an amateur astronomer can expect to observe.

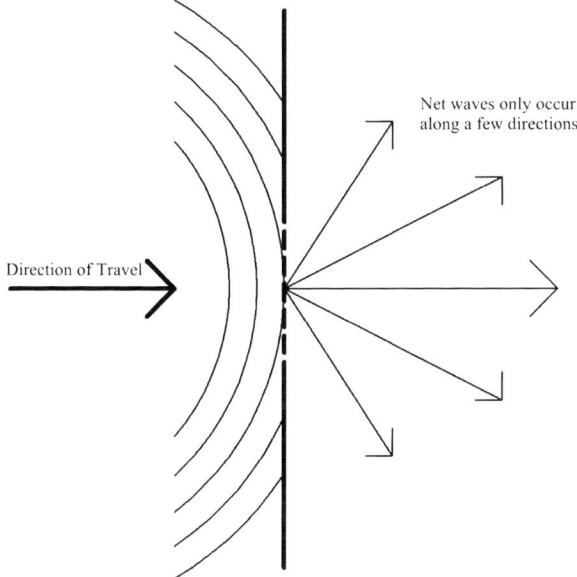

Fig. 8.19 When a wave traveling from *left* to *right* reaches a diffraction grating, net waves, only form by constructive interference in a few directions. The longer the wavelength, the greater the angle through which the wave "diffracts" (Image by the author)

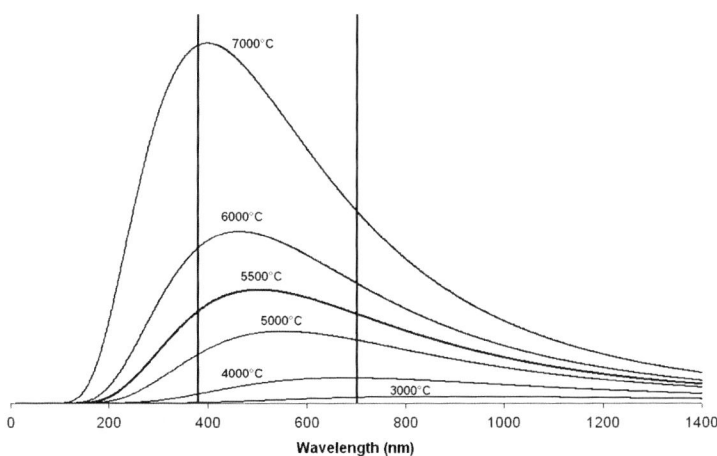

Fig. 8.20 Planck's law of black body radiation can be used to work out the temperature of the surface of a star from the amount of each wavelength present in the star's spectrum (ignoring *spectral lines*) (Image by the author)

Electrons in atoms and molecules can only possess certain energies. This curious fact puzzled scientists for many decades. No one got far in discovering why it is so until the atomic nucleus was discovered by Rutherford in 1911 [124]. Even though at this stage little was understood about the nature of the atomic nucleus – that knowledge came later – it soon became clear that most of the space in atoms and molecules is occupied by electrons.

Because of the wavelike nature of electrons (Chap. 3), it will be necessary to investigate briefly an analogy between electrons' wave functions and waves on a string.

Waves

Figure 8.21 shows a wave, or more accurately a snapshot of a wave, at some time that is arbitrarily labeled Time Zero, or Time 0. The wave is not static in time: it changes. Figures 8.22, 8.23, 8.24, 8.25, 8.26, 8.27, 8.28 and 8.29 show how this wave evolves over time.

If we were to follow the sequence from Time 0 to Time 8, the wave would return to its original position.

The wave is the same at time 8 as it was at time 0. It is said to have completed one *cycle*. In principle it will keep on cycling like this until something happens to it. The number of cycles the wave completes in one second is called the frequency of the wave.

The distance between two crests of the wave is called its wavelength (Fig. 8.30). In fact one wavelength is the distance between any two points in the wave having the same phase.

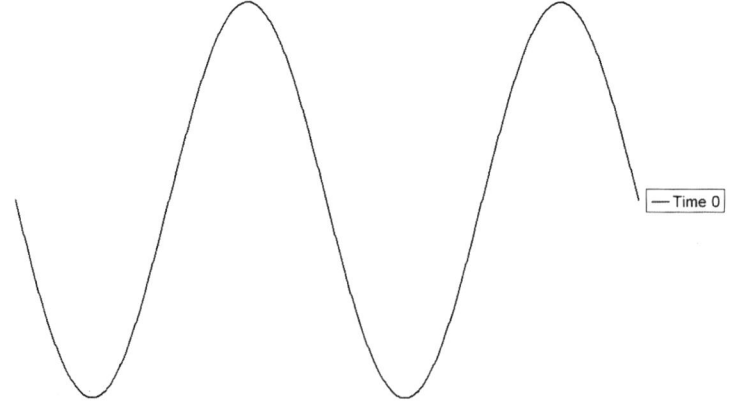

Fig. 8.21 An example of a wave, at time 0. Time zero is arbitrarily chosen (Image by the author)

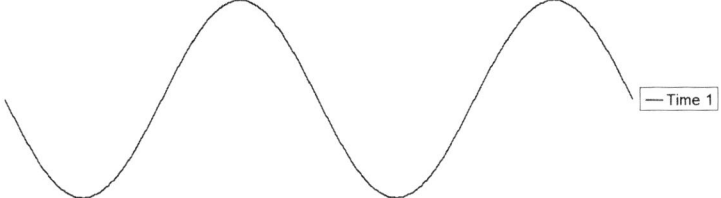

Fig. 8.22 The same wave, a little later, at time 1 (Image by the author)

Fig. 8.23 The same wave, a little later, at time 2 (Image by the author)

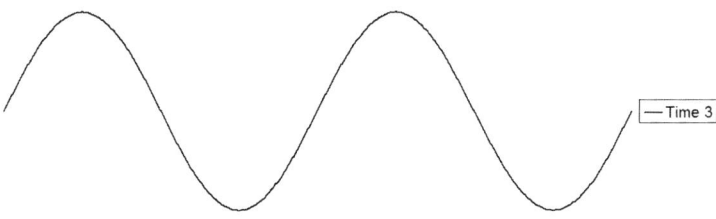

Fig. 8.24 The same wave, a little later, at time 3 (Image by the author)

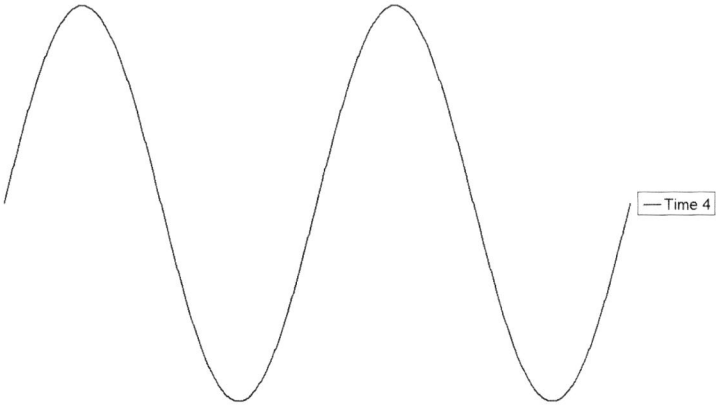

Fig. 8.25 The same wave, a little later, at time 4 (Image by the author)

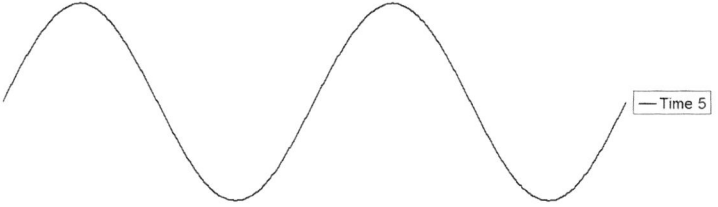

Fig. 8.26 The same wave, a little later, at time 5 (Image by the author)

Fig. 8.27 The same wave, a little later, at time 6 (Image by the author)

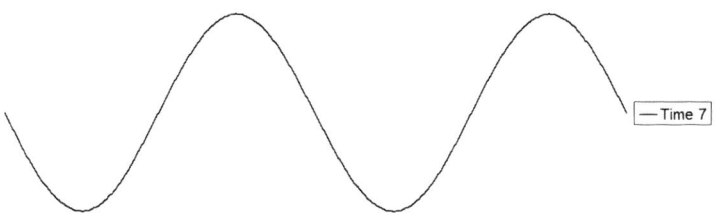

Fig. 8.28 The same wave, a little later, at time 7 (Image by the author)

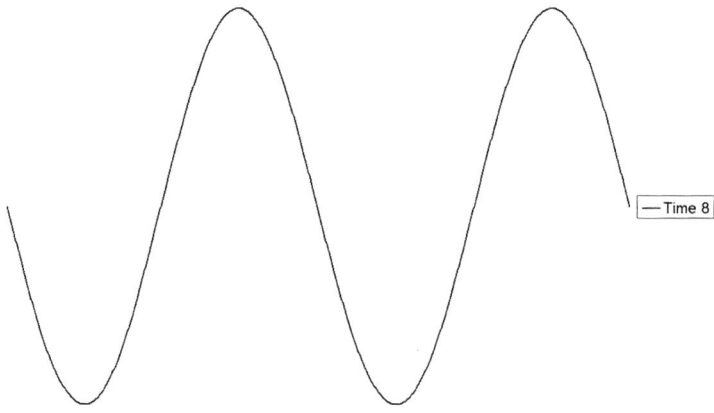

Fig. 8.29 The same wave, a little later, at time 8 (Image by the author)

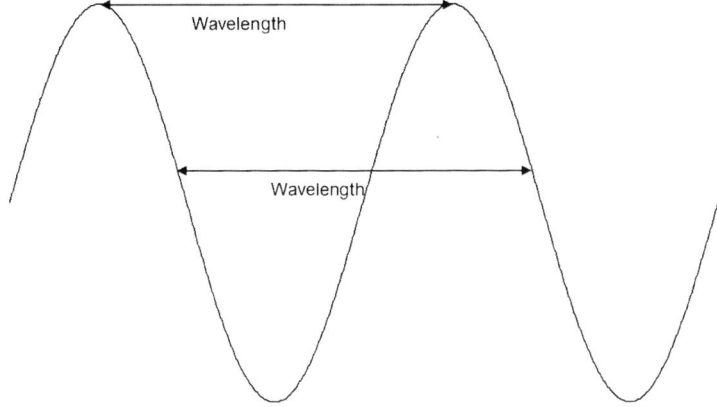

Fig. 8.30 The wavelength is the distance between two crests of the wave, or between any two points having the same phase (Image by the author)

Wave-Like Nature of Electrons

It was stated in Chap. 3 that particles like electrons can be thought of as point objects, and that the probability of finding this point object is given by entities called wave functions. These wave functions have a wavelength.

In practice these waves are not arranged in a straight line, as are the waves in the discussion above. They are arranged around the nucleus, which is a very tiny object at the center of the atom. They are arranged around the nucleus in three spatial dimensions. Let us consider what they might be like in two spatial dimensions for simplicity. In Fig. 8.31 a circular wave is shown in which exactly one wavelength fits around the circle. The wave does not look like what we are used to seeing as waves, but it is one.

In Fig. 8.32 a wave is shown whose wavelength is such that exactly two wavelengths fit around the circle. Why has the word "exactly" been used? This is because these waves have a remarkable property, which, as will be shown, is crucial to the phenomenon of spectral lines.

The property is that you can only have a whole number of wavelengths around the circle. You cannot have one-and-a-half wavelengths, for example. This is because the phase of the wave would no longer have a unique value at every point around the circle. Such a wave simply cannot exist.

You can, however, have waves with 3, 4, 5, 6, 7, 8 or more whole wavelengths going around the circle. Such waves are shown in Figs. 8.33, 8.34, 8.35, 8.36, 8.37 and 8.38. By the time you get to eight wavelengths around the circle (Fig. 8.38), the wave is beginning to look more recognizably wave-like, but all the waves shown are waves.

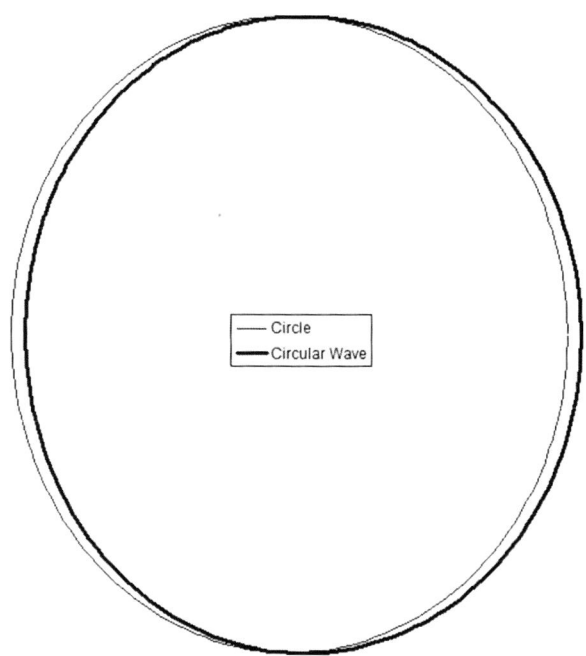

Fig. 8.31 A *circular wave* in which exactly one wavelength fits around the *circle* (Image by the author)

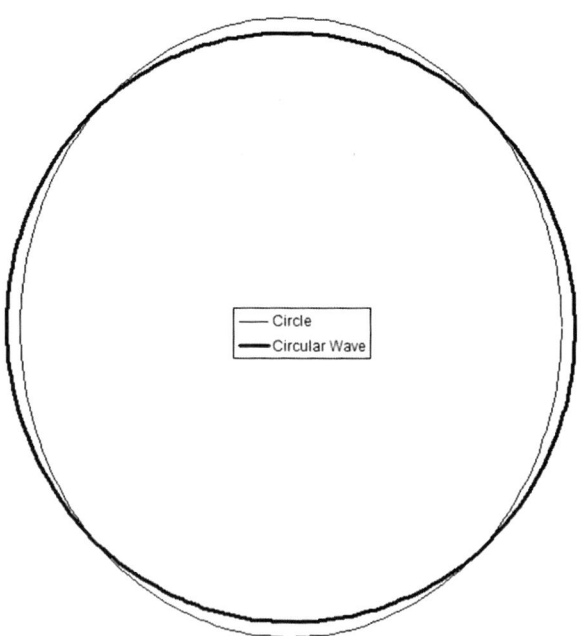

Fig. 8.32 A *circular wave* in which exactly two wavelengths fit around the *circle* (Image by the author)

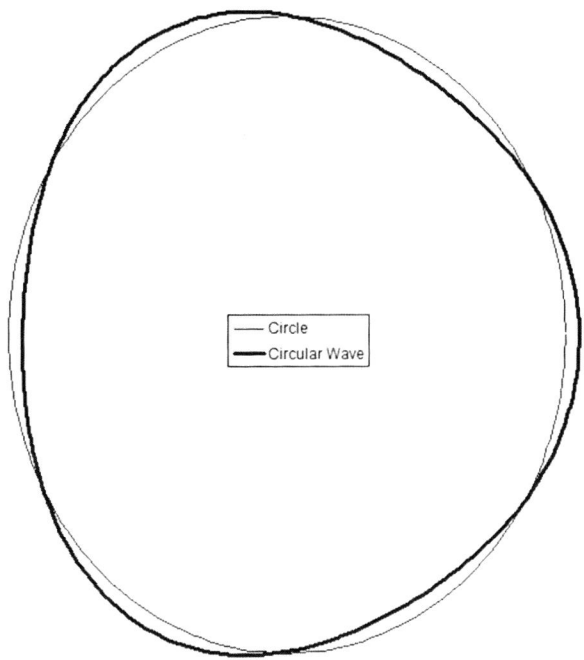

Fig. 8.33 A *circular wave* in which exactly three wavelengths fit around the *circle* (Image by the author)

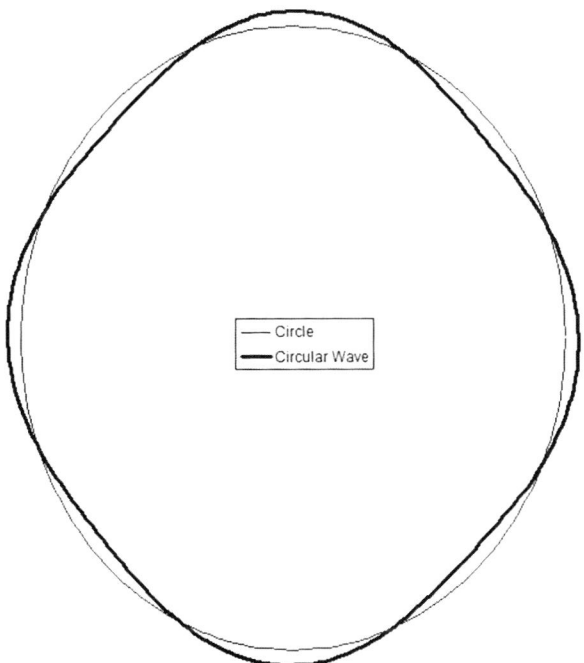

Fig. 8.34 A *circular wave* in which exactly four wavelengths fit around the *circle* (Image by the author)

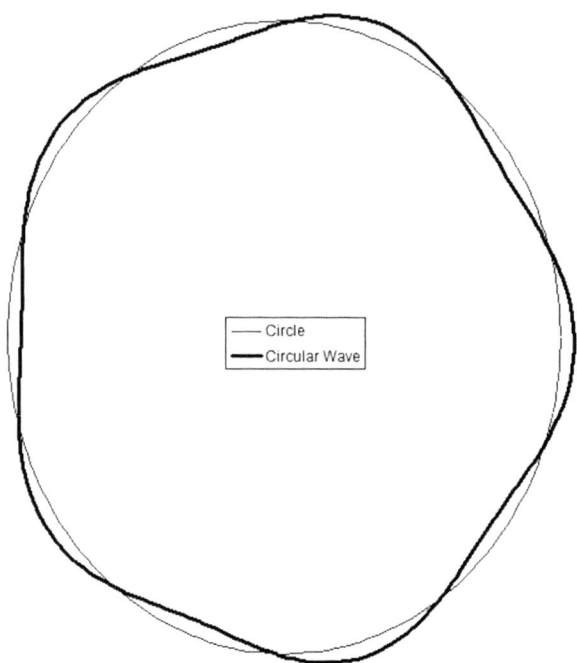

Fig. 8.35 A *circular wave* in which exactly five wavelengths fit around the *circle* (Image by the author)

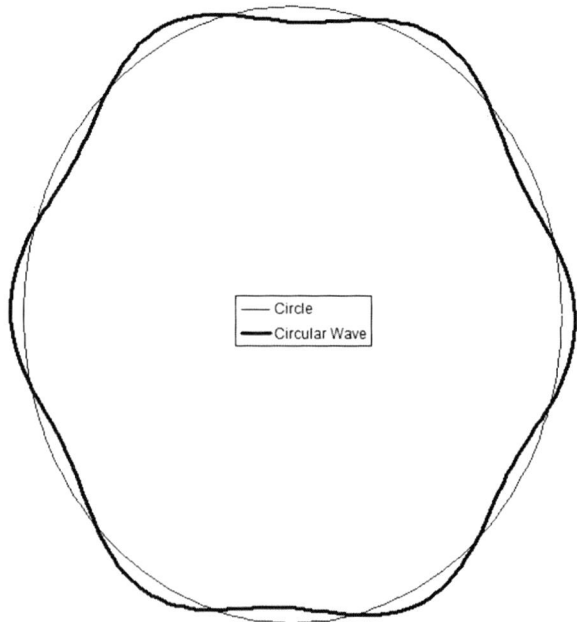

Fig. 8.36 A *circular wave* in which exactly six wavelengths fit around the *circle* (Image by the author)

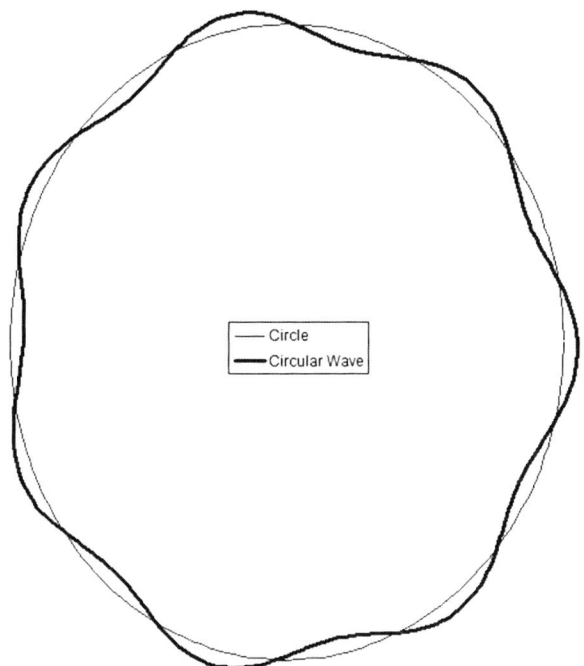

Fig. 8.37 A *circular wave* in which exactly seven wavelengths fit around the *circle* (Image by the author)

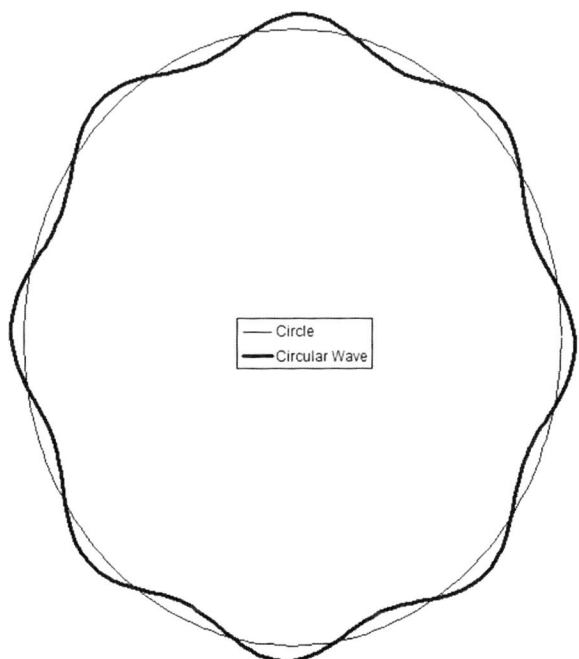

Fig. 8.38 A *circular wave* in which exactly eight wavelengths fit around the circle (Image by the author)

Generally speaking, the more wavelengths you fit around the circle, the higher the frequency of the wave will be. Although it is difficult to prove that this is always true, you can observe it for yourself with a rope such as a jump rope, known outside North America as a skipping rope. If two children turn this rope slowly, it will rotate with just one crest. If they turn it fast enough, they can set up two rotating crests, and a stationary point in the middle. By going even faster, they can get two stationary points. In other words, as they increase the frequency they change the number of wavelengths.

The geometry of the waves in three dimensions around a real atom is very complex. The mathematics required to describe such waves is positively eye-watering [125]. Nevertheless, the same thing happens to them. Only certain wavelengths can exist.

Where Photons Come into this Story

The speed at which a wave travels is simply the number of wavelengths that pass a given point in one second (or whatever your preferred unit of time is). It does not matter if it is a wave with a wavelength of 1 foot, and frequency one cycle per second, or a wave with a wavelength of one inch and frequency 12 cycles per second. They both travel at one foot per second.

Einstein discovered that the speed of light, at which photons travel, is always the same, approximately 1,86,000 miles per second or roughly 300,000 km per second [126].

Therefore as the wavelength of the photons varies, the frequency must vary to maintain the constant speed, which is popularly known as the speed of light, even though it is the speed of all photons, whether visible or not.

The early quantum physicists discovered that the energy associated with a given photon is equal to h times its frequency where h is always equal to 6.626×10^{-34} J seconds, or 0.000 000 000 000 000 000 000 000 000 000 000 626 J seconds. It's a very tiny number. If you are left none the wiser by the idea of a Joule second, don't worry. You do not need to know. The point is that if you know the energy of a photon, you know its frequency and vice versa.

You also know its wavelength because all photons travel at the same speed.

At Last, We Come to Spectral Lines

We have established that the wave functions governing the electrons in atoms can only have certain wavelengths. Therefore if you try to put energy into the atom, in the form of photons, to change the wavelengths of the electron's wave functions, the atoms will only absorb or emit those photons that change the wavelength by an allowed amount, viz. by an amount that leaves a whole number of wavelengths

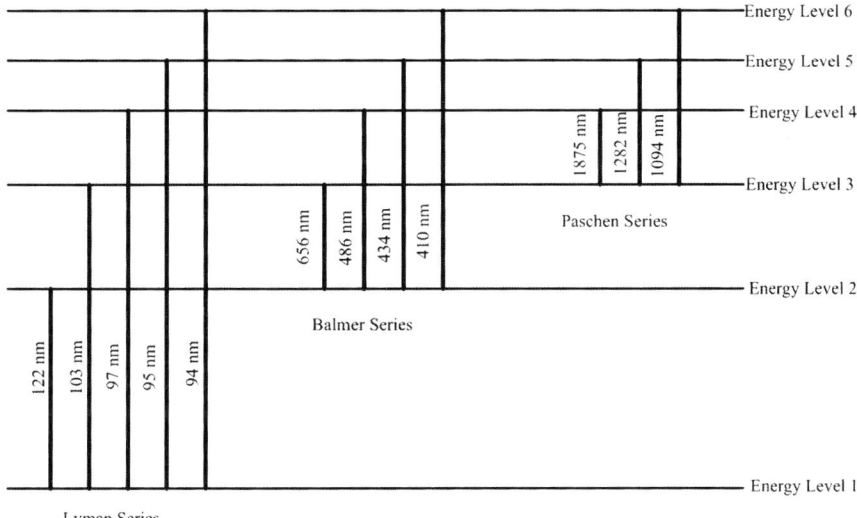

Fig. 8.39 The wavelengths of photons that will cause energy changes (i.e., wavelength changes) to an electron in a hydrogen atom [127]. The names of these series appear to be a little random because they pre-date our full understanding of what causes these spectral lines. This should come as no surprise – science is inherently experimental and observational. Theories usually come after observations, especially a revolutionary theory like quantum mechanics. Once the theoretical explanation has been developed, all sorts of apparently disparate phenomena are seen to be linked (Image by the author)

in the wave function. All other photons will sail right through the atom without being absorbed.

All right. If you want to be really picky, there is probably a teeny, tiny gravitational interaction, but this is way too small to be measured by any experiment scientists can carry out today. The account here also omits some subtle magnetic and relativistic effects.

Most of the atoms in the Sun are hydrogen atoms. The wavelengths of photons that these are allowed to absorb are shown in Fig. 8.39. Photons with these, and only these, wavelengths can interact with hydrogen atoms. Photons having any other wavelengths will not interact with hydrogen atoms, but instead will pass through them unscathed. If you look at all closely at the numbers in this diagram, you will see that the energy levels 4, 5 and 6 are very close together. This diagram is not to scale. There are many more than six energy levels. Only the first six have been shown.

It should be pointed out that the only stable state for an atom is that in which all electrons occupy their lowest possible energy level. Electrons occupying higher energy levels tend to return very rapidly to their lowest energy level very quickly. In the process they throw out a photon. This emitted photon might be sent

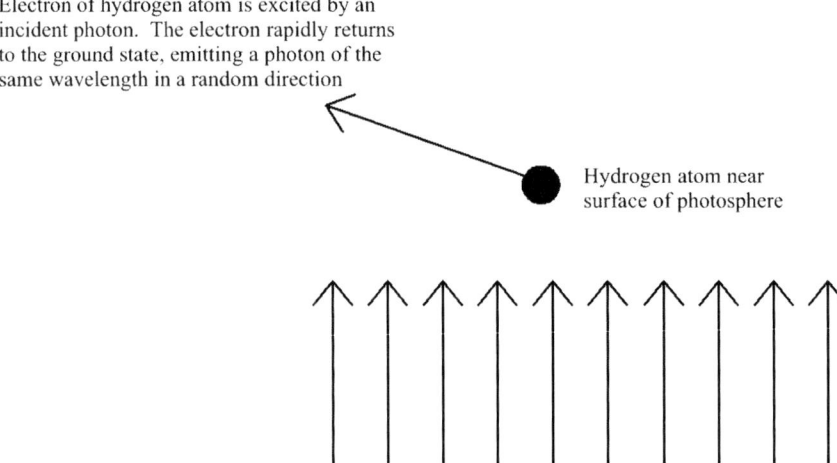

Electron of hydrogen atom is excited by an incident photon. The electron rapidly returns to the ground state, emitting a photon of the same wavelength in a random direction

Hydrogen atom near surface of photosphere

Incident photons, from all directions but overwhelmingly from the center of the Sun

Fig. 8.40 The origin of absorption spectral lines. If photons of a wavelength that can interact with an atom in their path do so, the photons are absorbed, and then quickly re-emitted in a random direction (Image by the author)

off in any direction. The lowest possible energy state is called the ground state. Figure 8.40 shows such absorption and re-emission close to the surface of the Sun's photosphere.

The reason some wavelengths are darker in the spectrum of a star is that the photons of that wavelength have found atoms with which they can interact and have been scattered as shown in Fig. 8.40.

What Can an Amateur Expect to Achieve?

There is a small but thriving amateur astro-spectroscopy world. Available equipment falls into four categories: simple home-made, low resolution, mid resolution and high resolution.

The lowest-tech approach is that the markings on a blank compact disc (CD) or DVD are small enough and close enough to act as a reflective diffraction grating. You can easily make such a device yourself. Figure 8.41 gives the 'recipe.'

How good is this spectrograph? It will not show the dark absorption lines in the solar spectrum to the naked eye, but it certainly shows that the solar spectrum is continuous, and redder than the spectrum from a cloudy sky. It is possible to

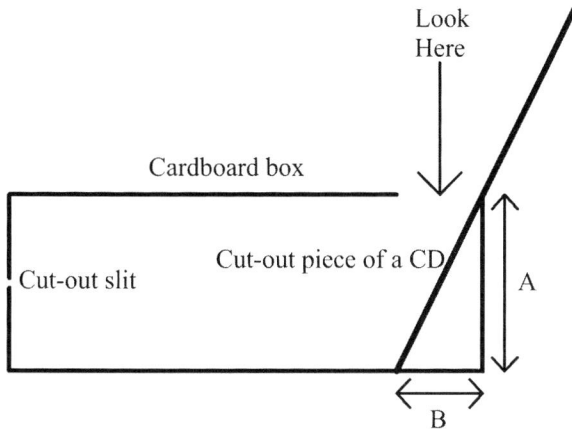

Fig. 8.41 A do-it-yourself spectrograph made with a box such as a cereal box and a CD or DVD. Length *A* should be about twice length *B*. If you look at the Sun with this device, (**a**) do not use any lenses other than your eyeglasses and (**b**) use your common sense and be careful. It is not safe to stare at the Sun, even with the naked eye. The diffracted spectra are much dimmer (Image by the author)

photograph the spectrum produced. Once this is done, the image should be converted to monochrome in *Photoshop*™ or a similar package. If the image is then sharpened in *Registax,* the D and F lines of Fig. 8.15 can be seen. This technique is excellent at showing that incandescent light bulbs have a continuous spectrum, whereas energy-saving and fluorescent light bulbs do not. In other words, frankly, as far as astronomy is concerned, it's a toy (Figs. 8.42 and 8.43).

The next level up is a transmission diffraction grating in a 1.25-in. filter holder. This filter shown in Fig. 8.44 will enable you to compare the spectra of the Sun and other stars as well as getting information about reflected spectra from planets.

To obtain an image like that shown in Fig. 8.45 requires the use of a software package such as *RSpec,* www.rspec-astro.com, which is downloadable for a fee.

RSpec will work live with an image from a webcam or DSLR and give immediate feedback about whether your exposure is correct.

Ironically, you will get better results photographing spectra with a monochrome camera. This is because color cameras work by splitting every set of four pixels into two red, one blue and one green (Fig. 8.46). First, you get more resolution if the filters are absent. Second, the analysis software does not have to worry about the different intensities of light transmitted by the pixel filters.

These gratings are designed to take point sources of light and analyze them. An extended source like the Sun will not work directly. You will need to use some ingenuity to get a point-like sample of sunlight into your camera. One option is to place a piece of black paper with a pinhole over, or behind, the grating, but you must then also use a white light solar filter such as the ones shown in Figs. 8.3 or 8.8. The Sun is so bright that you will have plenty of available light with which to do this.

Fig. 8.42 A spectrograph like that shown in Fig. 8.41. In practice it proved to be easier to observe the spectrum of the Sun from indoors, in a room where the blinds had been mostly pulled down over the window. If this photograph is in color, the spectral lines are quite invisible. If the color is removed, two lines can be seen. Better results were obtained with a blank CD than with a blank DVD (Image by the author)

Fig. 8.43 Close-up of the spectrum shown in Fig. 8.42, enhanced with Registax, showing the D and F lines of Fig. 8.15 (Image by the author)

Fig. 8.44 The Star Analyzer 100 from Patton Hawksley Education Ltd of England. This is a transmission grating with 100 lines per millimeter, mounted in a 1.25-in. filter holder (Image courtesy Patton Hawksley, used with permission)

The next level of sophistication might be an instrument like Elliott Instruments' new CCDSPEC, see for example http://www.elliott-instruments.co.uk/ccdspec.pdf. This instrument can resolve spectra down to the nearest 1.5 nm (Fig. 8.47).

This instrument comes with its own analysis software, *PCSpectra.* The manual shows that it is capable of analyzing the spectra of planets, thereby showing which components of sunlight are absorbed, which in turn gives clues to the atmospheric chemistry of the planets.

This last capability comes about because it is equipped with a slit so that only a slice of an extended object needs to be displayed. Through a telescope, Solar System planets are of course extended objects.

Elliott Instruments have announced a new version, which will use a reflection diffraction grating instead of a prism, and which will have a resolution of 1 nm.

The manufacturer warns against pointing the telescope holding the spectrometer against the Sun. You could try this with a filter such as the ones shown in Fig. 8.8, but do this with the greatest care, and do not blame anyone but yourself if you fry your equipment (Fig. 8.48).

Other instruments with higher specifications exist, but they are so powerful, and resolve so finely, that they need top quality cooled CCD cameras to take advantage of them. Hence they are beyond the reach of all but the wealthiest amateurs.

Solar Magnetism

At this point, it should be mentioned that, although the above discussion omitted magnetic effects, magnetism can be detected via spectral lines. The theoretical explanation of this effect would detain us for literally years. It is only skated over in undergraduate physics majors, but the experimental discovery is plain enough.

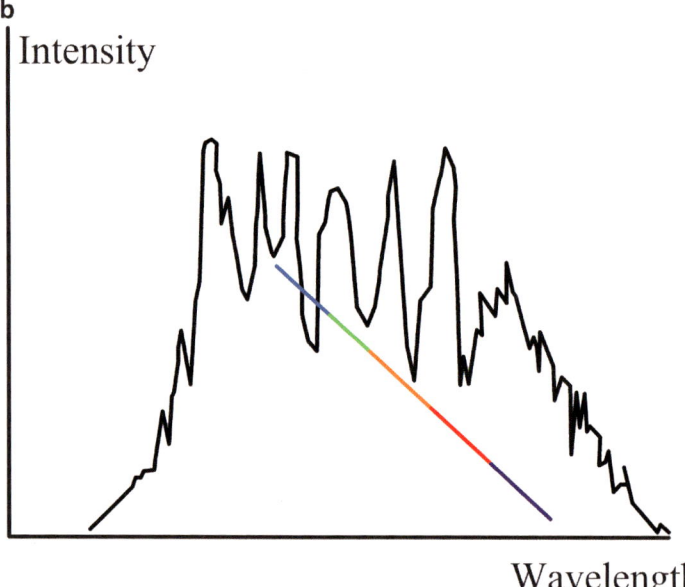

Fig. 8.45 The star analyzer shown in Fig. 8.44 claims to turn an image like the one on the *left* into an image like the one on the *right*. You need to buy a software package to produce the graph (Images courtesy Patton Hawksley, used with permission)

In 1896, Pieter Zeeman discovered that if he created the D lines of Fig. 8.15, which are due to the chemical element sodium, in a laboratory, they would split into several lines if the sodium were placed in a strong magnetic field [128]. One of his photographs is shown in Fig. 8.49. Since this was not the task his boss had set him,

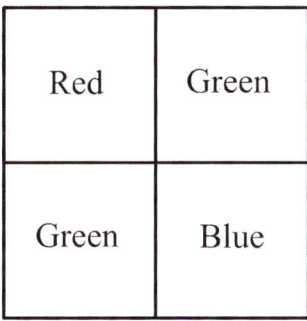

Red	Green
Green	Blue

Fig. 8.46 In a color camera, filters are placed over the pixels as shown. These filters are not present in monochrome cameras, which therefore have more resolution for a given pixel density (Image by the author)

Fig. 8.47 The Elliott Instruments CCDSPEC spectrometer attached to the back of a catadioptric telescope with a Meade DSI II camera attached (Image from Elliott Instruments, Ltd., used with permission)

he was fired for his pains, but he got the last laugh. The 1902 Nobel Prize in Physics was his for this discovery, known now as the Zeeman effect.

Beginning in 1908, George Ellery Hale, he for whom the Hale Telescope was named, discovered that the Zeeman effect occurs in spectral lines at sunspots [129]. Subsequent work by Hale and co-workers established that the polarity of sunspots is aligned east–west, that the polarity is opposite in opposite hemispheres, and that the magnetic polarity switches in each hemisphere from one sunspot cycle to the next. A photograph by Hale and colleagues is shown in Fig. 8.50.

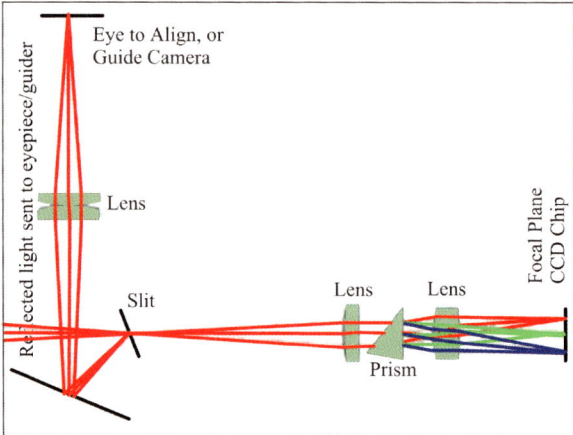

Fig. 8.48 Schematic diagram showing how the Elliott Instruments CCDSPEC spectrometer works. It uses a prism, not a diffraction grating (Image from Elliott Instruments Ltd., used with permission, annotation by author)

Fig. 8.49 Zeeman discovered that the D spectral lines of Fig. 8.15, when created in a laboratory and subjected to a strong magnetic field, would split into several lines (Image courtesy of Pieter Zeeman. Nature, vol. 55, 11 February 1897, pg. 347, Translated by Arthur Stanton from the Proceedings of the Physical Society of Berlin)

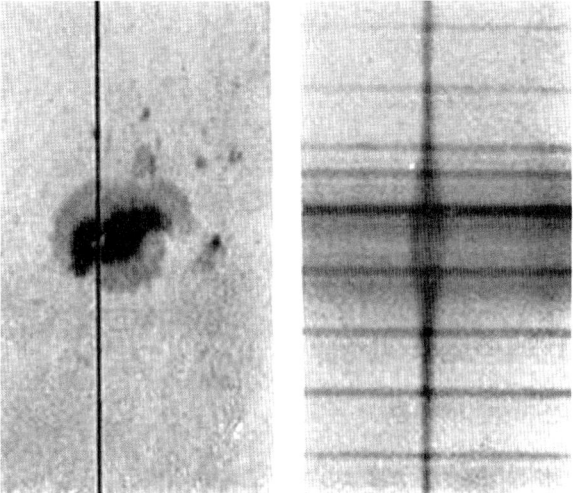

Fig. 8.50 The Zeeman effect observed in a spectral line at a sunspot (Image by Hale and others [130])

In a later paper [131], Hale explains how polarity at sunspots is determined because when spectral lines are split by the Zeeman effect, the two lines into which a line splits are oppositely polarized. In Hale's day Nicol prisms were used to polarize light. Nowadays, filters can be purchased with Polaroid sheets in them. It is observed in laboratories that if the magnetic field giving rise to the Zeeman effect is reversed, the polarity of the lines into which the original splits also reverses. This is how Hale and co-workers determined the magnetic polarity of sunspots.

One of the more interesting aspects of reading the papers of the great astronomers is the care that they took to make sure that they really were observing what they claimed, and to eliminate alternative possibilities such as thermal expansion in the spectrometer. This shines through Hale's writings. He wrote beautifully – not for him dense, turgid prose.

Could you as an amateur check the magnetic polarity of sunspots? It certainly is not a commonly reported amateur feat. Hale reports Zeeman splitting wavelengths of 0.01–0.05 nm. The resolution of the Sheylak Lhires III spectrograph is not quite adequate to pick this up. Packaged spectrographs that are available are designed for looking at faint stars, not for looking at the Sun. If you wanted to increase the resolution, the trick is to buy a good diffraction grating and make your own spectrograph that magnifies the spectrum more than the commercial ones do. Maurice Gavin reports building a prism-based spectroscope in the 1980s [132] that displayed a solar spectrum 40 in. long. In those days photographic film was the medium, and you can buy gratings separately from firms like Shelyak. So there is a challenge for someone feeling bold.

A theoretical explanation of the behavior of sunspots took a long time. Arguably, we are still not fully there. Unfortunately for most readers, this subject is very mathematical indeed. The mathematically adventurous could bring themselves up to speed by reading the book [133] and a short course article [134] by Arnab Rai Choudhuri. It is interesting to note that in the latter, a prediction was made that sunspot cycle 24, which is due to peak in 2013–2014, would be weak. This prediction has certainly been borne out. A more recent Internet article suggests that other authors have not been so successful in sunspot cycle intensity predictions [135].

Helioseismology: A Tool for Observing Solar Structure

The internal structure of Earth has been determined by analyzing the propagation of seismic waves through the planet. We cannot listen to the Sun in the same way because there is no atmosphere to transmit solar sounds our way. There is a very thin plasma in the Solar System. It is not a total vacuum between the planets. But this medium is too thin to transmit anything detectable.

Cunning is required instead. Seismic waves propagate rather slowly through the Sun. In principle, the movement of the Sun as the waves propagate should produce a Doppler effect (Chap. 5, Fig. 5.2) in the spectral lines observed. The change in the Doppler effect would be visible on a timescale of days.

It is therefore inconvenient that the Sun rises and sets once a day. There are various ways around this problem, such as having a network of observing stations at various longitudes, observing from the South Pole where the summer daylight lasts 24 h, or observing from space. These have all been tried during the 50 years that helioseismology has been practiced.

Helioseismology will only be mentioned in passing, as it is not an easily accessible subject for amateur astronomers. The analysis of data obtained is highly mathematical, and the Doppler "shifts" in spectral lines are too fine for amateurs to detect [136]. Indeed some of this work is done from spacecraft such as the Solar Dynamics Observatory, launched in 2010, which is in a geosynchronous orbit at 102°W, inclined at 28.5° to the equator. Thus it spends most of its time over the eastern Pacific Ocean.

Helioseismology can detect things inside the Sun. We amateurs can only look at its surface. One of the most impressive results is that we now know how fast the Sun rotates throughout its volume.

Figure 8.51 shows the rotation period at different depths and latitudes. At the surface, it is possible to observe visually by following sunspots that the solar rotation varies from about 25 days at the equator to about 31 days at 60° north or south. Helioseismology reveals that this differential rotation only happens in about the outer 20 % by radius of the Sun. The innermost 60 % by radius rotates at a

Rotation
Period
(Days)

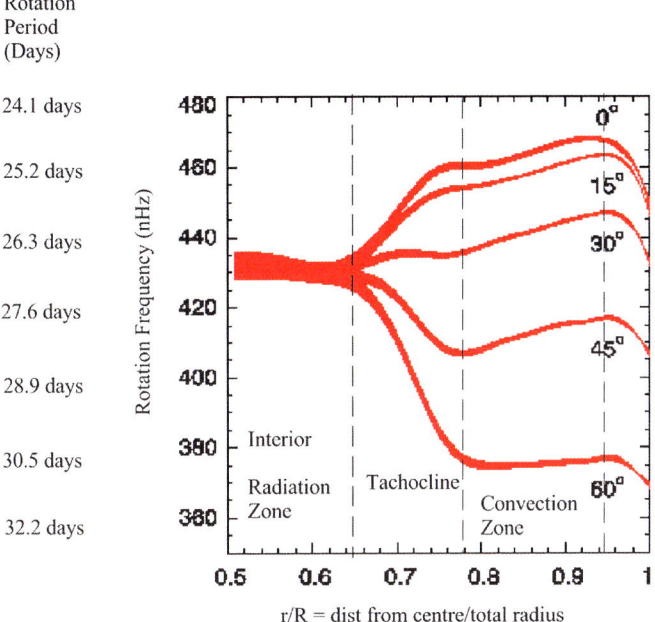

Fig. 8.51 The internal rotation of the Sun about its axis. The horizontal axis is the fractional radius: r/R=0.5 means halfway from the center to the surface, r/R=0.9 mean 90 % of the way from the center to the surface, etc. The vertical axis is rather inconveniently given in nanoHertz – billionths of a cycle per second. At the far left, these units have been converted into a rotation period in days. The innermost 65 % by radius rotates at a fairly constant rate of about once every 27 days. At the surface, the rotation period is about 25 days at 0° latitude and about 31 days at 60° latitude (Photograph courtesy of Global Oscillation Network Group, licensed under Creative Commons Attribution-Share Alike 3.0 Unported license. http://en.wikipedia.org/wiki/Helioseismology#mediaviewer/File:Tachocline.gif)

constant rate, taking about 27 days. These two regions are joined by a region called the *tachocline*.

We know from other theoretical studies that the radiation zone broadly occupies the constant-rotation region, and the convection zone broadly occupies the differential-rotation region [137].

Chapter 9

Small Fry: Asteroids and Comets

Compared to the planets, these objects are mere specks of dust. The largest by far, Ceres, has about 1/300th of the volume of Earth. The others in the 'big four' asteroids, Pallas, Vesta and Hygieia, have about a tenth the volume of Ceres [138]. Between them they contain most of the mass in the Asteroid Belt. The Hubble telescope can resolve Ceres into a sphere and has measured its rotation period. Hubble-era values for the sizes of the asteroids [139] are noticeably different from those quoted in the 1977 *Cambridge Encyclopedia of Astronomy* [140].

Ceres is the only one of the four whose gravity has pulled it into a roughly spherical shape. Vesta and Pallas are sort of spherical, but only sort of. A NASA space mission, in part funded by Germany and Italy, called Dawn, is now in orbit around Vesta. It is due to fly on to Ceres, arriving there in 2015 [141].

There is a strong consensus that the asteroids are protoplanets, the bodies that gravitationally captured one another to form the planets we know today, although they may have histories of agglomeration followed by collisions [142]. The reason they never coalesced into one or more planets is generally believed to be that Jupiter's gravitational pull prevented this from happening. Indeed, the asteroids are clearly shepherded in their orbits by Jupiter (Fig. 9.1) [143]. There is another factor in all this: the combined mass of all the asteroids is tiny compared to that of a rocky planet. A more plausible (to the author) hypothesis therefore is that the asteroids are protoplanet remnants that Jupiter's gravity prevented from getting sucked into other planets.

© Springer Science+Business Media New York 2015
J. Clark, *Viewing and Imaging the Solar System*, The Patrick Moore
Practical Astronomy Series, DOI 10.1007/978-1-4614-5179-2_9

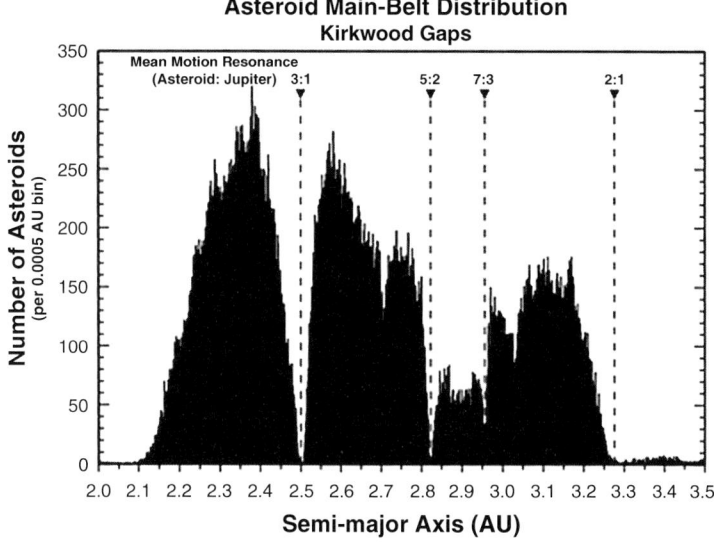

Fig. 9.1 There are few or no asteroids in the 'Kirkwood gaps' where the orbital periods are in a simple ratio with that of Jupiter (Image courtesy of NASA)

The argument that they would have made a separate planet seems to come down to the Titus-Bode law. This is a rule giving the distances from the Sun to the planets. It has a gap between Mars and Jupiter, which worked for Uranus but did not predict the correct distance from the Sun to Neptune. Everybody got excited when Ceres seemed to be the 'missing planet' in Bode's law. Murray and Dermot [144] show that the chances of a law just as good as the Titus-Bode law arising by chance is very high. In other words, the Titus-Bode law is spurious, and no significance should be attached to it. It is not a reasonable argument for saying that there is a 'missing' planet between Mars and Jupiter.

So that's what an asteroid is – a leftover planetesimal orbiting the Sun but guided by Jupiter. Let's have a look at what you can expect to observe. With amateur equipment, you are only ever going to see asteroids as dots. The only way to tell them from stars is to track their movements.

The photographs shown in Figs. 9.2 and 9.3 track the path of Ceres through its 2008–2009 apparition. In each case Ceres is circled. You can see that it has moved noticeably in a day. The only way to identify it was to look for an object whose position had changed.

The orbital elements change on a timescale of a decade or so, because of Jupiter's pull. Orbit tables in packages such as *Cartes du Ciel* therefore need to be updated. To do this, manually look up the orbital elements at http://www.minorplanetcenter.net/iau/MPCORB/MPCORB.DAT and edit the text file *Cartes du Ciel* uses. Looking in the catalog menu enables you to find this file. It takes ages for this file to load, so please be patient.

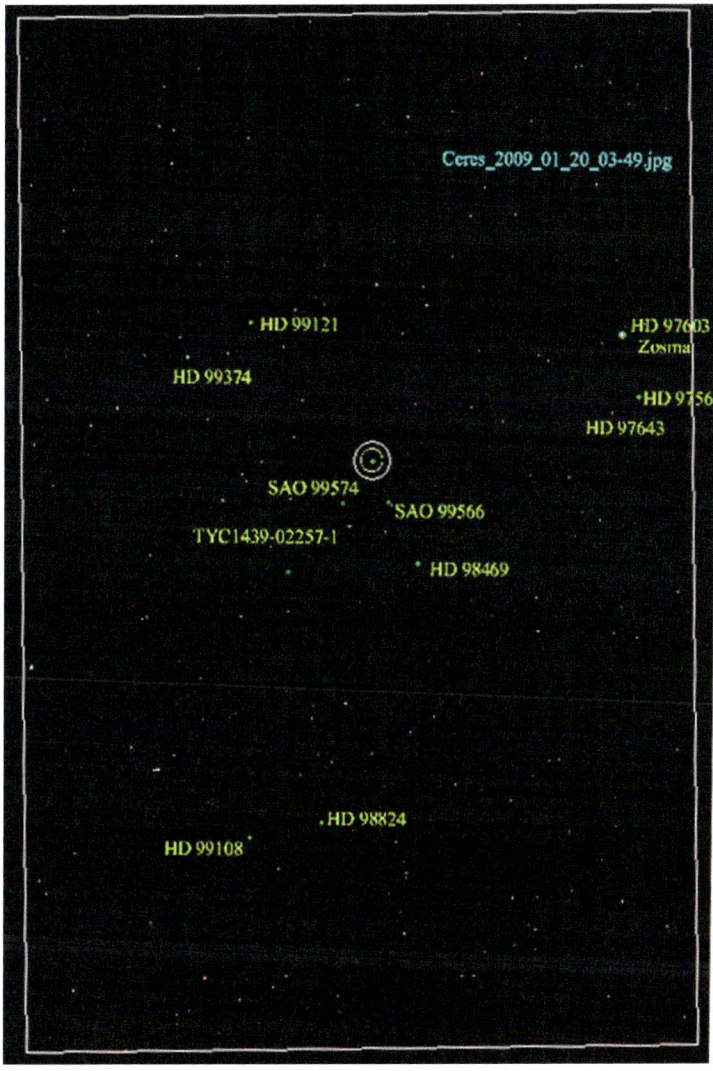

Fig. 9.2 Ceres, January 20, 2009, 03:49 UT. Taken with a DSLR, on a driven equatorial mount, unguided, 200 mm lens, 30 s at f/5.6. Ceres is circled (Image by the author)

The path of Ceres through Leo during the 2008–2009 apparition is shown in Fig. 9.4.

In Fig. 9.5, the motions of Ceres and Saturn during two apparitions are compared. Ceres' orbit is tilted at only 10° to Earth's orbital plane (the ecliptic). It follows that the orbit of Saturn must be almost coplanar with that of Earth. We see hardly any end-on movement, whereas we are somewhat looking down onto the plane of Ceres' orbit.

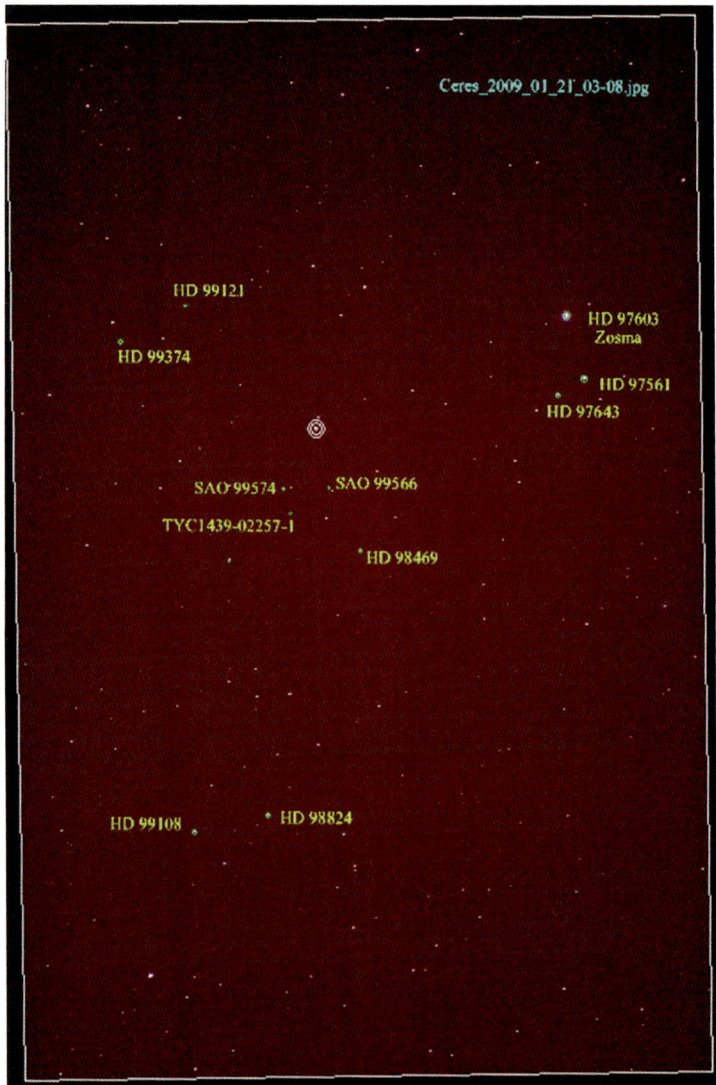

Fig. 9.3 Ceres, January 21, 2009, 03:08 UT. Taken with a DSLR, on a driven equatorial mount, unguided, 200 mm lens, 30 s at f/5.6. Ceres is *circled* (Image by the author)

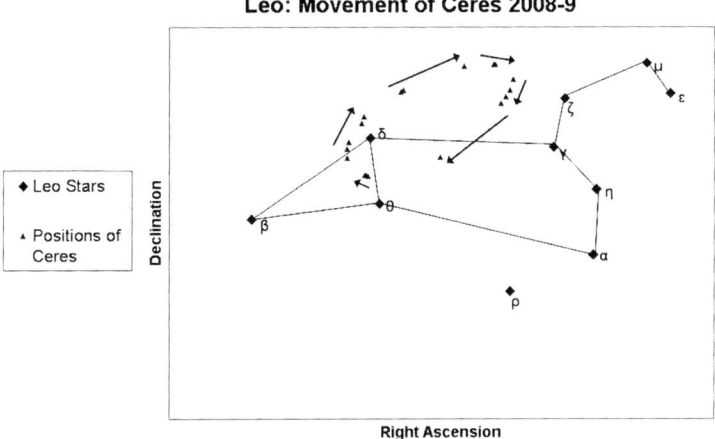

Fig. 9.4 This shows how Ceres moved through the constellation of Leo during the 2008–2009 apparition (Image by the author)

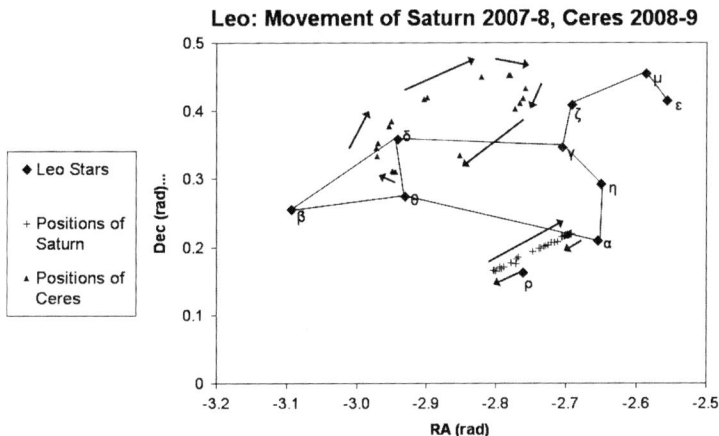

Fig. 9.5 Comparison of motions of Ceres during the 2008–2009 apparition, and Saturn during the 2007–2008 apparition (Image by the author)

Comets

It is hard for us to believe now what fear comets struck into people only a few generations ago. The only great comet of the author's lifetime – Hale-Bopp in 1996–1997 – came and went without much popular fear. The last big naked-eye one in 1910 caused all sorts of scares [145], not least that by this time the scientists were

Fig. 9.6 Comet Shoemaker-Levy smashes into Jupiter in 1994, leaving 'splash marks,' some bigger than Earth (Image courtesy of NASA)

able to detect toxic organic materials in the tail, and the press never seem to miss an opportunity for a good old pseudo-scientific scare story. Unscrupulous entrepreneurs sold people 'comet pills' to combat the 'threat' of this gas.

There was a very bright comet in 1811. Tolstoy, writing about the Napoleonic invasion of Russia half a century later in *War and Peace* [146], describes how it was said to portend "all kinds of woes and the end of the world."

The climate of fear around comets may have been part of the reason Isaac Newton devoted so much space in his great mechanics book [147] to deducing their orbits. These orbits tend to be very long, narrow ellipses [148].

What is a comet? It is an object made up of a nucleus a few miles across, made of silicate minerals and 'ices' – water ice, solid methane, solid ammonia and solid carbon dioxide. As they swing around the sun, the ices and CO_2 in comets evaporate and are broken up chemically, forming compounds that fluoresce in the sunlight. This is the material of the tail or tails [149]. (Hale-Bopp very noticeably had two tails in 1996–1997. You could easily see them in binoculars.)

In 1994 a comet called Shoemaker-Levy 9, which actually had several nuclei, crashed into Jupiter [150]. This event brought into sharp relief two ideas. One is that Jupiter acts as a comet cleaner-upper for the inner Solar System, helping to stabilize Earth as an environment suitable for life. The second is that if that comet had hit Earth instead of Jupiter, it would have been very nasty event indeed, making a nuclear war seem puny (Fig. 9.6).

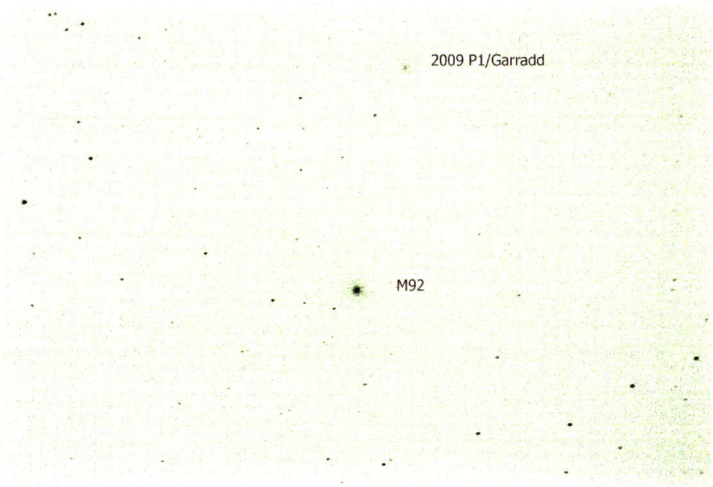

Fig. 9.7 Comet 2009 P1/Garradd as it passed M92 as seen from Earth. Taken with a Canon EOS 1000 DSLR, 500 mm catadioptric lens, 120 s @ f/8 (Image by the author)

Bright comets are not so common. There are always dim ones about. The SOHO solar-orbiting observatory [151] publishes images, from which amateur astronomers have more than doubled the number of comets for which orbits have been determined since Isaac Newton showed us how. Unfortunately, you are not going to see these dim ones from your backyard. Indeed at any one time there are only one or two around that are claimed to be binocular objects. These are usually very hard to find. In general, to find comets, we are talking about looking up right ascension and declinations on websites [152, 153] and using the telescope's GOTO capabilities.

It proved to be impossible to image 17/P Holmes with a webcam in 2007, even though it was easily visible in binoculars. Photographing comets requires deep sky methods.

Figure 9.7 shows a photograph of seventh magnitude comet 2009 P1/Garradd using a piggybacked DSLR. Also visible in the image is the globular cluster M92, which has visual magnitude of 6.3. The comet is much less bright in this image than the globular cluster. This suggests that the magnitude of 7 claimed for the comet by the British Astronomical Association for the comet may be too high. For example, the star immediately to the bottom left of M92 is HD156821, whose magnitude is 9.76 according to *Cartes du Ciel*. It is brighter than the comet. The comet certainly was not visible in 7×50 binoculars, which a seventh magnitude object ought to be.

The nomenclature is that 2009 is the year of discovery. P means 'periodic' (as opposed to C, 'non-periodic'; X, 'orbit could not be determined'; D, 'disappeared' or A, 'oops it was really an asteroid'.) Garradd is the name of the discoverer.

Fig. 9.8 Comet 103P Hartley, taken with the Faulkes robotic telescope on October 15, 2010, at 12:34 UT (Image by Isabella Hayward, used with permission)

Just to make sure you are thoroughly confused, there are other systems of comet nomenclature.

The Faulkes Telescope project (http://www.faulkes-telescope.com/aboutus) is a project involving two robotic telescopes, one in Hawaii and one in Australia, which are made available free of charge to schools and to astronomical societies working with schools. The project is supported by the Dill Faulkes Educational Trust.

The telescopes are 2-m (79-in.) f/10 Ritchey-Chrétiens. Figures 9.8, 9.9, 9.10, 9.11, 9.12, 9.13, 9.14 and 9.15 are images of Comet 103P Hartley taken from Clifton High School in Bristol, England, by a student remotely using the Faulkes Telescope North in Hawaii. Exposure times were about 7 s.

In Fig. 9.14, the images of Figs. 9.8, 9.9, 9.10, 9.11, 9.12 and 9.13 are superimposed with the stars in the same position. From this image it can be deduced that the rate of the apparent motion of the comet through the sky is about 7.68 arc sec per minute of time.

Fig. 9.9 Comet 103P Hartley, taken with the Faulkes robotic telescope on October 15, 2010, at 12:36 UT (Image by Isabella Hayward, used with permission)

In Fig. 9.15, the six images have been stacked on the comet using *Deep Sky Stacker* (Chap. 4). The stacking software appears to have dropped two images and stacked four. The most striking thing about this stacked image is how much more non-nuclear comet material is visible than in the individual images. As with most digital photography of Solar System objects the moral is clear: you see more if you stack.

Further information on the observation of comets can be found in the companion book in this series by James and North [154].

Fig. 9.10 Comet 103P Hartley, taken with the Faulkes robotic telescope on October 15, 2010, at 12:38 UT (Image by Isabella Hayward, used with permission)

Fig. 9.11 Comet 103P Hartley, taken with the Faulkes robotic telescope on October 15, 2010, at 12:40 UT (Image by Isabella Hayward, used with permission)

Fig. 9.12 Comet 103P Hartley, taken with the Faulkes robotic telescope on October 15, 2010, at 12:41 UT (Image by Isabella Hayward, used with permission)

Fig. 9.13 Comet 103P Hartley, taken with the Faulkes robotic telescope on October 15, 2010, at 12:42 UT (Image by Isabella Hayward, used with permission)

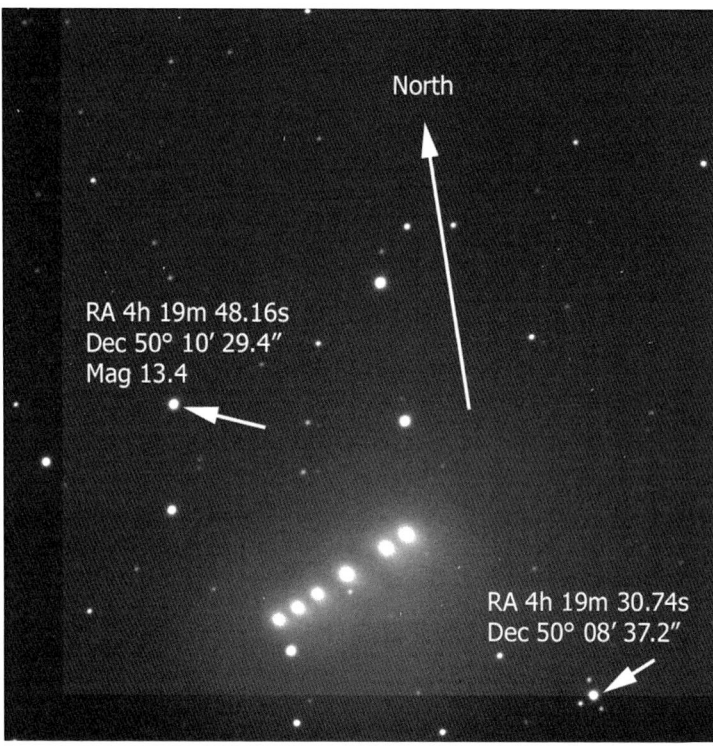

Fig. 9.14 Comet 103P Hartley, taken with the Faulkes robotic telescope on October 15, 2010. 9.8 through 9.14, keeping the stars fixed at 12:42 UT (Image by Isabella Hayward, used with permission. Annotations by the author)

Fig. 9.15 Comet 103P Hartley, taken with the Faulkes robotic telescope on October 15, 2010, at 12:42 UT (Image by Isabella Hayward, used with permission, and annotations by the author)

Chapter 10

The Apps and Downs of Mobile Devices for Astronomy

Introducing the Geek Speak

If you are old enough, you may remember a household device of 20 years ago called a video cassette recorder. Your memory of it will also depend on your age. If you were well into adulthood, you used to have to get your kids to work these devices for you because they were fiddly and not user friendly. If you were a child, you no doubt remember smugly how confused your parents were. This would have begun the fastest transition known to humanity – the transition from parents knowing everything in the eyes of their children to knowing nothing in their eyes.

Nowadays the touchscreen-operated smart phone serves that function. Youngsters use them as if they were born knowing how to, while those older and wiser are frustrated and baffled by them. Like VHS video recorders they have small and fiddly controls.

So, you ask, what is a smart phone? You could think of it as a pocket computer that also makes phone calls. The technology is rapidly advancing, and no doubt before we all know it, today's smart phones are going to look primitive.

Most of this chapter also applies to tablets, which have nothing to do with pills, legal or otherwise. Rather, they are portable computers. Tablets will easily fit in a briefcase. There are also mini-tablets, which will fit conveniently into all but the smallest handbags.

A typical smart phone is shown in Fig. 10.1. A typical tablet is shown in Fig. 10.2.

They also give rise to an argot. If you are of a certain age, perhaps you will just about have gotten your head around the idea that computer people describe the

© Springer Science+Business Media New York 2015   215
J. Clark, *Viewing and Imaging the Solar System*, The Patrick Moore
Practical Astronomy Series, DOI 10.1007/978-1-4614-5179-2_10

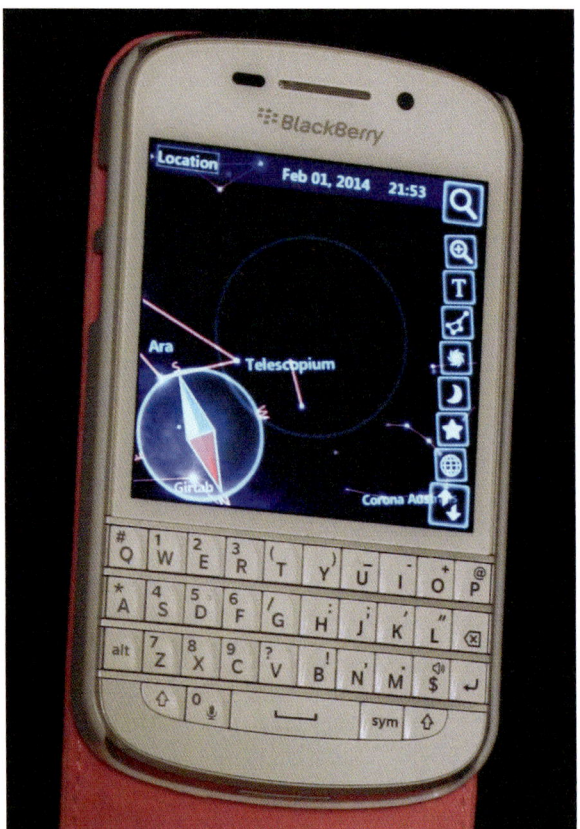

Fig. 10.1 An example of a smart phone (the BlackBerry Q10). The external appearance of such devices is usually very similar, except that most smart phones are touch-screen only. The App being shown in this case is 'Star Tracker' (Image by the author)

software you load onto your PC as 'applications' of the PC.' They are no longer 'programs,' even the slightly less antediluvian 'software packages.' On phones and tablets, programs are called 'apps.'

The other thing you have to get your head around is the concept of a smart tag. Smart tags are a bit like barcodes, only they are square. They are also called QR codes. QR stands for 'quick response.' Believe it or not, they were invented by Toyota so that they could track vehicles and components in their factories. Nowadays the cameras in smart phones and tablets can be used to read these tags. They can even create them. Figure 10.3 shows what one looks like.

Smart phones and tablets are typically operated through touch screens. Some have separate keypads, such as the BlackBerry Q10, but most do not. The author's experience is that it takes some time to get used to using touch screens. Once you master this art, you wonder why you ever had a problem. It needs to be mastered

Fig. 10.2 An example of a tablet (the iPad Mini). The app being shown is 'SkySafari' (Image by the author)

Fig. 10.3 An example of a QR code [156]. (Image courtesy of Autopilot, licensed under the Creative Commons Attribution-Share Alike 3.0 Unported license, http://en.wikipedia.org/wiki/File:Qr-3.png)

again if after having become accustomed to a tablet, you acquire a smart phone. Patience is the watchword. You may frequently have felt like heaving your desktop PC out of the window in frustration. What saved your PC from this unkind fate was the difficulty of accomplishing a defenestration. Smart phones are all too easy to hurl. A recent British prime minister was noted for his ability to cause his phones to fly [155]. It is not hard to see his point. They *are* exasperating little demons.

Several operating systems are in use on smart phones and tablets. The variety of operating systems is of more than academic interest. Many astronomy apps are only available on certain systems. You therefore need to have at least a vague idea what kind of smart phone or tablet you have. Android from Google appears on such phones as Nexus. BlackBerry has its own proprietary system. Its smartest phones currently operate on BlackBerry 10. Microsoft has a version of Windows for smart phones. Nokia, Samsung and other makers currently use it. At the time of this writing, the smart phone and tablet markets are dominated by the Apple operating system iOS, which only appears on Apple's own hardware.

All you have to do to get an app into your phone is to tap the QR code icon on your phone to read the tag using the camera in the phone, and follow your nose through the incomprehensible icons, which are clearly designed to save you, the reader, the trouble of reading the English language. Suddenly, hey presto! You are being offered a chance to buy the app.

The prices of apps are normally quite low. This no doubt is to keep their cost below the level at which piracy would be widespread.

Planetarium Apps

There is no shortage of these. They exploit the fact that smart phones and tablets generally contain magnetic compasses and can pick up global positioning satellite (GPS) signals. These devices therefore know the date, time and their locations. They also have built-in gyroscopes, digital compasses and accelerometers. They therefore know which way you are pointing them and how much you are moving them. They also have Wi-Fi™ (i.e., wireless Internet connection) and often other means of connection. The author's smart phone can connect via Wi-Fi, the BlackBerry network and the Vodafone network. (Vodafone is a cellular telephone network in the United Kingdom.) This limits the amount of information that you need to tell the app. It knows where it is, when it is and which way you are pointing the device.

Typical planetarium apps enable you to point your device at a celestial object and read from the screen what it is. It takes a little patience to understand exactly which way you are pointing the device: appearances can be slightly deceptive. You can zoom in the usual way of such devices by moving your index finger and thumb together or apart. All this is easy enough with a little practice.

One aspect of these apps is that bifocal and varifocal eyeglass wearers cannot read their screens if they point their devices upwards. This problem is relatively easily addressed in the software. All you should need to do is touch a screen icon to suspend the movement of the screen as you move the phone or tablet, to enable you to move it to where it would be visible through the close-up part of your glasses. At least one app, *Star Tracker* can do this, but it does not advertise that it can. You simply touch the screen, and it stops following the sky for long enough for you to read it.

There are more planetarium apps than you can shake a stick at. Most of them are not very sophisticated. They have a database of stars, 'fuzzies' such as galaxies and nebulae, and of course Solar System objects. You should look for an app with a night vision mode, in other words, a mode where the display is red; otherwise your device will prevent you from seeing the heavens themselves. You should be able to turn object names on and off; and if you are using the device to enable you to star-hop your way to a faint object, you do not want the app to supply a huge picture of the object.

The one that stands out from the crowd is *SkySafari*. It has three versions, at different price levels.

The basic version is much like other planetarium packages, except that if you touch a celestial object, it offers more information about the object than most apps. It has a database of 119,000 stars, the major planets and about 500 asteroids.

The 'Plus' version also includes telescope control, by wire or wirelessly. By this it means that you can use the app to make the telescope go to any object you select. It has a much larger database: 2.6 million stars, 31,000 deep sky objects, and more relevantly to us, 18,000 comets, asteroids and satellites. It can also show you the view from elsewhere in the Solar System.

The 'Pro' version has all the features of the plus version. In addition it has 27 million stars, three quarters of a million galaxies and more than 620,000 Solar System objects. It also has a Moon map based on NASA's Lunar Reconnaissance Orbiter data. It claims to show the sky with sub-arc second precision from anywhere on Earth, in the Solar System, or beyond, at any time up to one million years in the past or future.

Why might a Solar System observer care about stars? They are occasionally useful. For example, Jupiter and Saturn sometimes move near stars of similar brightness to their satellites. It is then useful and fun to be able to distinguish satellites from stars. An example is given in Fig. 10.4. More impressively, *SkySafari* also knew the position of the Great Red Spot when the photo shown in Fig. 1.5 was taken (Fig. 10.5).

Can *SkySafari* also predict the orientation of Saturn's rings? Comparison with Figs. 7.37, 7.38, and 7.39 shows that it can. It was noticed that if the time settings are displayed, you have to be careful not to move your finger over the "now" icon as you zoom in to look at the rings and zoom out to find Saturn as you change the date.

These feats proved to be impossible with *Star Tracker* on a BlackBerry Q10.

Fig. 10.4 Jupiter, four Moons and a star. The *upper* picture was taken on September 25, 2008, at 18:54 UT. The *lower* was taken almost exactly a day later on September 26, 2008, at 19:03 UT. The star is HR7128, apparent magnitude 5.92. It is about 640 light years away. It is a B8 *blue-white* star, with absolute magnitude −0.53 and has 138 times the luminosity of the Sun. In early 2014 it proved to be not only possible but easy to find all this information from SkySafari. Pictures taken with an 8-in. f/6 Newtonian, in prime focus, with a Philips SPC900NC webcam (Image by the author)

Fig. 10.5 A 'control' image was taken on September 26, 2008, at 18:52 UT to obtain the orientation of Jupiter. At this exposure level, none of the satellites nor the star seen in Fig. 10.4 is visible. The Great Red Spot is visible to the *lower right* of Jupiter's disk. SkySafari predicts its position well if we assume that the time displayed by the app is British summertime, not Universal Time. Picture taken with an 8-in. f/6 Newtonian, in prime focus, with a Philips SPC900NC webcam (Image by the author)

Other Solar System Apps

There are *dozens* of these. They are far too numerous to list without being boring. You can get children's introductions to the Solar System, atlases of the solid Solar System bodies from Horsham Online Ltd., JupiterMoons and SaturnMoons from Sky & Telescope Media LLC (both of which have night vision modes), and force and gravity simulations from Ravindar Kompella and from IL&FS Education & Technology Services Ltd. Depressingly, but not surprisingly, there are even more astrology apps.

The app world is your oyster.

Chapter 11

Observing the Solar System from Your Armchair

NASA puts much of its space exploration data on the Internet, copyright free. Some military stuff is doubtless kept under wraps, but that is also off topic. Since the folks at NASA have been very busy bees since the dawn of space exploration, there is an awful lot of NASA material available.

The purpose of this chapter is to show you where you can find stuff about the Solar System, especially images, but also scientific results. The material indicated is that which is likely to appeal to amateur astronomers whether they are trained scientists or not.

Another online gold mine, aimed at scientists, is the Astronomy Abstract Service, hosted by Harvard University at http://adsabs.harvard.edu/abstract_service.html. The abstracts of just about every technical paper on astronomy are stored here in searchable form. Usually a link to the scientific paper that has been abstracted is also there, but you may have to pay for the paper. One way into this database is to start with Wikipedia, and then follow the links in the references of whatever articles you start with. After that, follow your nose, going from reference to reference. It can be very rewarding to check out the original papers, because quite often they tell you about all sorts of wrinkles and complications not mentioned in encyclopedias. Although many papers are not for the faint of heart, a surprising number can be read by those with little or no scientific training.

The Hubble Space Telescope (HST) is a catadioptric telescope. The original had a Ritchie-Chrétien optical train until it was realized that, due to a manufacturing boo-boo, a corrective lens (actually a corrective mirror) was needed. Its optics are

© Springer Science+Business Media New York 2015   223
J. Clark, *Viewing and Imaging the Solar System*, The Patrick Moore
Practical Astronomy Series, DOI 10.1007/978-1-4614-5179-2_11

probably therefore unique and will never be replicated. The instruments on board nowadays make the correction themselves. The original correcting mirror has been removed and is now in a museum [157].

The HST is in a nearly circular orbit 370 miles above Earth, at an angle of 28.5° to Earth's equator. This angle is not coincidental: Cape Canaveral is at 28.5°N. The near-Earth orbit was chosen to enable space shuttle missions to service the telescope. It has the disadvantage that Earth can get in the way of observations.

Even though the Hubble has taken some wonderful pictures of Solar System objects (available at NASA online), these do not really play to its strengths. You could probably get better pictures from Earth-bound telescopes with adaptive optics. The real advantage of being above the atmosphere is that this telescope can photograph the faintest, furthest objects. This is not a great advantage for taking pictures of the planets. In the cases of Mars, Jupiter and Saturn, there is probably a public relations aspect to the photos, since spacecraft orbiting these planets can do better by virtue of their proximity. The optics were corrected during the first servicing mission in 1993. In the years that followed, there had been no Mars Reconnaissance Orbiter and no Cassini/Huygens orbiter around Saturn. The Galileo orbiter arrived at Jupiter in 1995. In the 1990s, except for Jupiter, the Hubble photos of planets were state of the art.

One of the more curious phenomena that became clear from NASA pictures was that in visible light, the surface of Jupiter is full of swirling storms, but that of Saturn is not. Why would that be, when the two planets are somewhat similar? If Saturn is photographed at infrared wavelengths, the storms can be seen in the photographs.

Even if you are a keen outdoor astronomer, then, armchair astronomy can teach you stuff (Fig. 11.1).

The website for the Hubble telescope is www.hubblesite.org. You can drill down into this site (that is, by clicking on links). You will learn a lot about the telescope and its instruments, all in non-mathematical language.

There are many wonderful Solar System photos on this website. Figure 11.2 shows the limit of what Hubble can photograph. Although the image of Io is very impressive compared to the white dot you see in most amateur telescopes, it is not in the same league as the images sent back by spacecraft that visited Jupiter. In February 2007, *Cartes du Ciel* works out that Jupiter was almost 5.7 AU from Earth, or about 530 million miles away. You would need a really large telescope to outperform a local one.

The Sun

The SOHO probe, already mentioned, contains a wealth of solar images, located at http://sohowww.nascom.nasa.gov. The visible images of the Sun, such as Fig. 8.2, are by no means the only data available. There are two sets of coronal images, C2 and C3, showing the corona out to 5.25 million miles and 30 million miles from the solar center respectively.

Fig. 11.1 This shows how Saturn has storms like Jupiter, but that they only manifest themselves in infrared images (Image courtesy of NASA and the Space Science Telescope Institute. Source: http://hubblesite.org/gallery/tours/tour-saturn/fullscreen. This figure is a 'still' screen dump from a movie)

Then there are the X-ray images. These pick out very high temperatures. The solar atmosphere is hotter than the photosphere [158]. By going to higher and higher temperatures you are therefore looking higher and higher into the solar atmosphere.

Figure 11.3 shows X-ray images at four wavelengths. Going from left to right these correspond to temperatures of 1 million Kelvin, 1.5 million Kelvin, 2 million Kelvin and 60–68,000 Kelvin. The temperature in Kelvin (usually abbreviated K) is measured from absolute zero. A temperature change of 1 K is the same as a temperature change of 1 °C or 1.8 °F. Since in degrees Celsius, absolute zero is at −273.16 °C, the temperature in Kelvin is equal to the temperature in Celsius plus 273.16° (either Celsius or Kelvin). When compared to a million degrees, a difference of just under 300° really is neither here nor there. To a first approximation, you can read very high temperatures in Kelvin as being the temperature in degrees Celsius.

Finally, there are two sets of MDI, or Michelson Doppler Interferometer, images: a color image taken at a wavelength of 676.8 nm, and a magnetogram, which shows sunspots and similar features, with black and white against a gray background indicating opposite magnetic polarities.

It is worth noting that in Fig. 11.4, the leading sunspots in the pairs have opposite polarities in opposite hemispheres of the Sun.

All these images are archived, and the archives are easily searched. An example is given in Fig. 11.5.

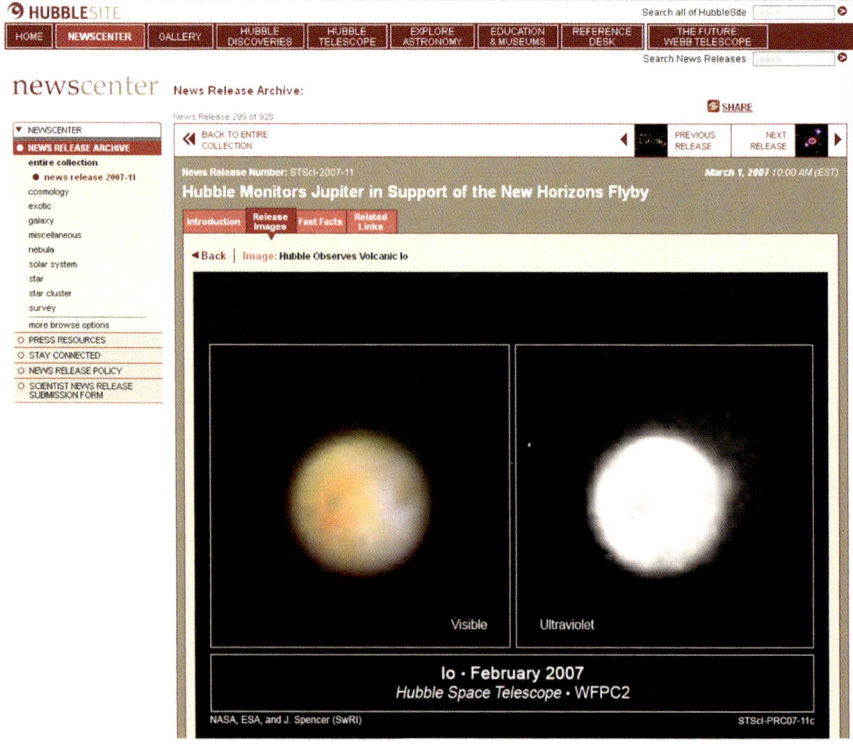

Fig. 11.2 This pair of images of Io shows the limit of what the Hubble can see in the Solar System. The space probes that went to Jupiter sent back much more detailed images. The ultraviolet image shows volcanic activity (Image courtesy of NASA. Source: http://hubblesite.org/newscenter/archive/releases/2007/11/image/f/format/web_print)

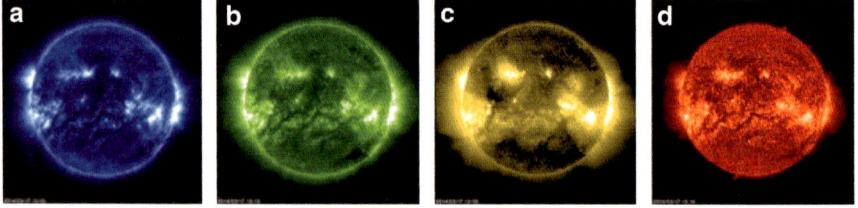

Fig. 11.3 SOHO images at various wavelengths, 17.1 nm, 19.5 nm, 28.4 nm and 30.4 nm, going from left to right respectively. These wavelengths are all in the X-ray range and might be compared to wavelengths of visible light from 380 nm to 770 nm (Images courtesy of NASA. Source: http://sohowww.nascom.nasa.gov/data/realtime/image-description.html)

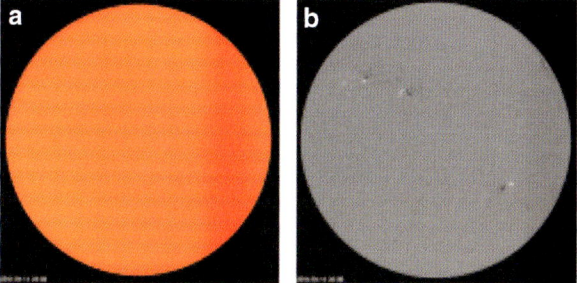

Fig. 11.4 MDI (Michelson Doppler Imager) images (Images courtesy of NASA. Source: http://sohowww.nascom.nasa.gov/data/realtime/image-description.html)

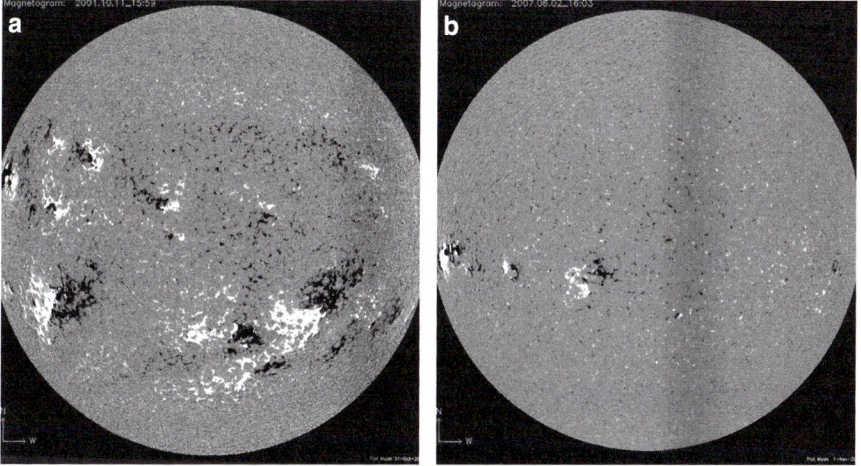

Fig. 11.5 Two magnetograms chosen at random from the SOHO archive. The left hand one, from November 11, 2001, is near the maximum time of sunspot activity. The right hand one, from June 2, 2007, is near the minimum time of sunspot activity (Source: http://soi.stanford.edu/production/mag_gifs.html)

Mercury

The first of two space probes to visit Mercury was *Mariner 10*, in the mid-1970s. It first flew by Venus, to obtain a gravity assist, and then flew by Mercury three times in a highly elliptical orbit, which it is presumably still following, although after three flybys the spacecraft was no longer functioning. Of course, in those days, the Internet was not even a twinkle in Tim Berners-Lee's eye, so there was no uploading of pictures very soon after they were taken. Instead, there is a complete book of the *Mariner 10* project at http://history.nasa.gov/SP-424/sp424.htm. The link to the pdf version of the book appears to be dead, but the HTML version (i.e., the normal Internet version) works fine. This mission did not manage to image the whole of Mercury. The images that are available can be found online at an Arizona State University website: http://ser.sese.asu.edu/M10/IMAGE_ARCHIVE/ Mercury_search.html.

It was mentioned earlier that the Messenger probe is in orbit around Mercury at the time of this writing. This spacecraft has on board two CCD cameras for visual imaging, a gamma ray spectrometer that can detect certain chemical elements up to 4 in. below the surface, a neutron spectrometer that sees hydrogen (chemically bonded in minerals) at up to 16 in. below the surface, an X-ray spectrometer that again sees certain chemical elements but has no depth penetration, a magnetometer, a laser altimeter, a spectrometer to determine the composition of the very tenuous atmosphere, and a radio device that measures Mercury's gravity by making use of the spacecraft's positioning data. The project's website is http://messenger.jhuapl. edu/index.php. By following the links from here, it is possible to find out a great deal about this mission.

Venus

The challenge in observing Mercury from close up is getting there. The challenge with Venus is seeing through the atmosphere (Fig. 11.6).

The earliest missions to Venus were not successful. Eventually in 1966 the Soviets managed to get Venera to enter the atmosphere. It never reached the surface because the parachute had been designed before people knew how dense the Venusian atmosphere is, so it stopped transmitting part-way down. At about the same time, NASA managed a flyby with *Mariner 5*. This was at the height of the Cold War between NATO countries and the USSR. Nevertheless American and Soviet scientists collaborated to analyze their data.

Armed with the knowledge thus gained, and expecting the atmospheric pressure at the Venusian surface to be between 25 and 100 atmospheres, the Soviets launched *Venera 5* and *Venera 6* in 1969. These probes were designed to survive to 25 atmospheres and had smaller parachutes. They were not expected to survive to the surface. They didn't, but they each returned data for almost an hour.

Venera 7 was designed to withstand 180 atmospheres. Unfortunately it seems to have suffered a parachute malfunction. In October 1975, *Venera 9* and *Venera 10* sent back the first-ever images of the surface. Soviet perseverance was rewarded at last. There is a nice website describing the mission that brought back the first pictures of the Venusian surface at http://mentallandscape.com/V_ Lavochkin2.htm. We now know that the atmospheric pressure at the surface is something like 90 Earth atmospheres. This is so dense that parachutes are not needed for the final descents.

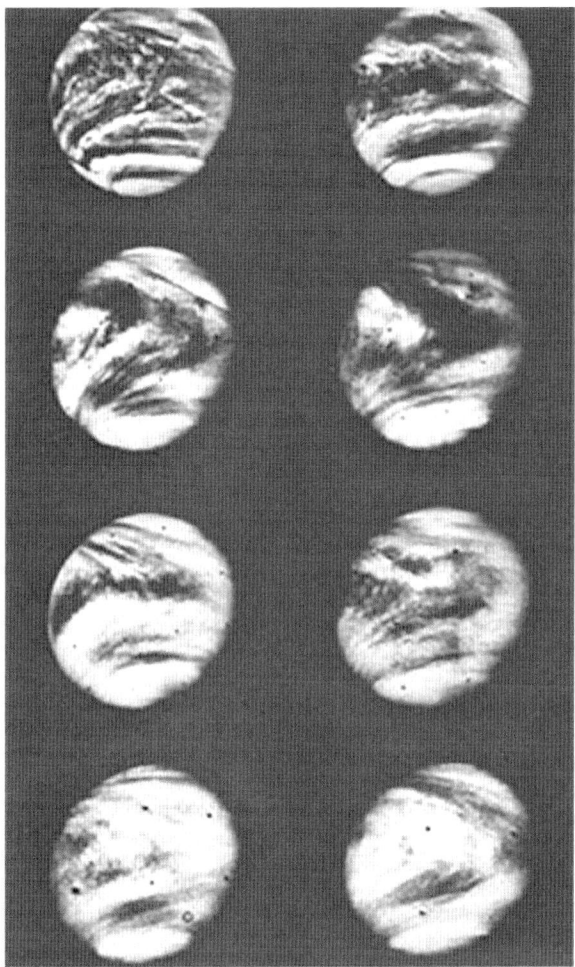

Fig. 11.6 Mariner 10 took this sequence of UV images of the clouds of Venus in February 1974. The resolution is better than that of more recent images from the Hubble telescope (Image courtesy of NASA. Source: http://history.nasa.gov/SP-424/ch6.htm)

By now the managers of the Venera program knew how to land successfully. One of the problems with photography was that the probes had to land on the daylight side, making them unable to transit radio signals to Earth. The landers had detached from platforms that flew by Venus and were able to transmit data for an hour or so before they flew out of range. Unfortunately the cameras on *Venera 11* and *Venera 12* (both 1978) failed, so no visual images were sent back. The most spectacular discovery made by these two probes was of lightning storms in the Venusian atmosphere. Resulting images can be found at http://mentallandscape. com/C_CatalogVenus.htm.

Two more Soviet probes, *Vega 1* and *Vega 2* sent landers to study the Venusian atmosphere. They arrived in June 1985. The main part of these spacecraft went on to Comet Halley.

Apart from flybys to hitch a ride to other planets, the remaining missions to Venus were orbiters: NASA's Magellan and the European Space Agency's Venus Express. Magellan arrived in 1990. It used radar to map the surface of Venus. The result can be visualized at http://laps.noaa.gov/albers/sos/venus/venus4/venus4_rgb_cyl_www.jpg.

Venus Express inserted itself into orbit around Venus in April 2006. Its results are less immediately available to a popular audience than those missions that take photographs. These results have been published in the scientific journal *Nature* [159].

Earth

Of course, most of the images we ever see of Earth are taken from its surface. After all, we live here. There are too many satellite images of Earth to mention them all. Google™ Maps, Google Earth, etc., enable you to call up a satellite image of almost any land area. Google Maps is based on a Mercator projection, so it cannot handle requests for satellite images of Antarctica, but these are not difficult to find elsewhere, e.g., at http://lima.usgs.gov.

The first Earthrise picture from *Apollo 8* is at http://www.hq.nasa.gov/office/pao/History/alsj/a410/AS8-13-2329HR.jpg. Wide publication of views such as this brought home to people just what kind of a world we live on and may have given a boost to environmentalist thinking. The iconic picture was the one from *Apollo 17* (Fig. 11.7).

This picture was remarkable when it was first published. Nowadays there are many unmanned satellites quite capable of taking such pictures (Fig. 11.8).

The Moon

How could anyone who lived through the late 1960s and early 1970s forget the Apollo missions to the Moon? People actually went there. So far these twelve men are the only people to have walked upon any surface other than Earth. There is a gallery of images at http://www.apolloarchive.com/apollo_gallery.html. You can spend many an hour clicking through these images.

The image shown in Fig. 11.9 is one of many very fine images brought back. It is hard not to suspect that someone took the trouble to train the astronauts in the art of taking good pictures.

Many other unmanned missions have taken place. NASA's Lunar Reconnaissance Orbiter entered a polar orbit 31 miles above the Moon in 2009, and it is still there at the time of this writing, early 2014. It found the remains of the manned missions

Fig. 11.7 The so-called "Blue Marble" picture of Earth, taken from Apollo 17 as it flew to the Moon (Image courtesy of NASA. Source: http://www.nasa.gov/images/content/115334main_image_feature_329_ys_full.jpg)

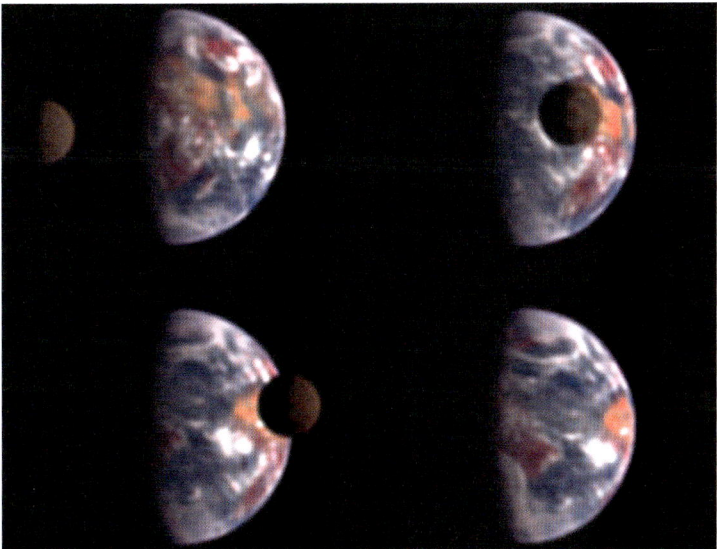

Fig. 11.8 Earth and the Moon, viewed from 31 million miles away by the Deep Impactor mission to Comet Tempel 1. These images form part of a sequence that you can view as a movie at http://www.nasa.gov/topics/solarsystem/features/epoxi_transit.html (Image courtesy of NASA)

(see, e.g., http://www.nasa.gov/content/lunar-reconnaissance-orbiter-looks-at-
apollo-12-surveyor-3-landing-sites). It is the first space probe to have the imaging
power to do this. The website for this probe is http://www.nasa.gov/mission_pages/
LRO/main/index.html#.UvgJDoXkc-g.

One of the more striking achievements of this mission was the production of
color relief maps of the Moon. There is such a map in the Wikimedia Commons at
http://upload.wikimedia.org/wikipedia/commons/b/b0/MoonTopoLOLA.png,
which was produced by a planetary scientist based in Paris, Mark A. Wieczorek,
whose home page is http://www.ipgp.fr/~wieczor (Fig. 11.10).

Mars

Even though no one has yet been to Mars, the surface of the Red Planet has been
crawled over more thoroughly than that of Earth's Moon, which humans have visited. This is partly because of the sheer hassle of getting people to and from Mars.
A nine-month journey there, followed by a similar journey back is a whole different
ball game than a week-and-a-half trip to the Moon. You need a lot more of everything: food, air, radiation shielding, and so on. Artificial gravity provided by cen-
trifugal action may also be necessary.

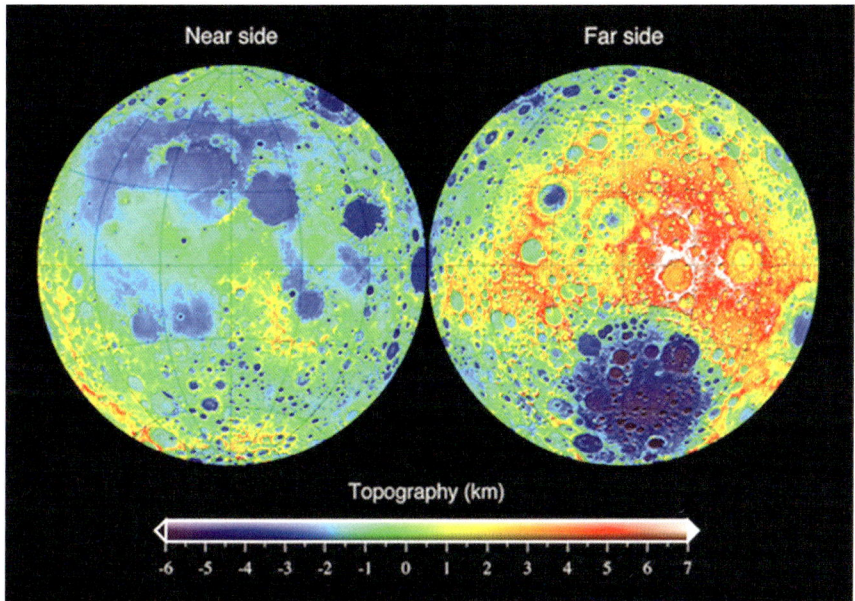

Fig. 11.10 Topography of the Moon, obtained from the Lunar Reconnaissance Orbiter (Image by Mark A. Wieczorek, 2010. Source: http://upload.wikimedia.org/wikipedia/commons/b/b0/MoonTopoLOLA.png)

Another factor is the remarkable improvement in vehicle quality since the Apollo project. The automotive industry has driven this, not least via the quality revolution that began in Japan [160]. Nowadays we feel quite irritated if our cars break down. Thirty years ago, frequent breakdowns were simply expected. The same applies to Mars rovers. We simply did not have the capability to produce space exploration vehicles that would last a decade in the Apollo era.

We certainly have this capability now. The Mars Exploration Rover program began when the first of two identical rovers, named Spirit, landed in January 2004. It was active until it broke down in 2010. It was sent to Crater Gusev [161]. Its twin, named Opportunity, landed also in January 2004, on the Meridiani Planum on the opposite side of the planet [162]. It is still in operation as of February 2014. Both were close to the Martian equator. They are illustrated in Fig. 11.11.

Many of the panoramic images sent back are available in stereo, e.g., http://marsrovers.jpl.nasa.gov/gallery/press/opportunity/20140214a/pia17943-FigB-SOL3527-McLure_L7R1ana.jpg. You need 3-D glasses of a rather old-fashioned type, with a left lens in red and a right lens in blue, to see these. If you are an eye-glass wearer, you can buy clip-on ones on eBay. Note that you do *not* want the ones with a green left lens and a magenta right lens. You also do not want the kind used to view modern television screens in 3-D. Also note that just occasionally, they publish the two colors the wrong way around, e.g., in the case of Astronomy Picture

Fig. 11.11 An artist's impression of one of the Mars exploration rovers, Spirit and Opportunity (Image courtesy of NASA. Source: http://marsrovers.jpl.nasa.gov/gallery/artwork/rover1 browse.html, NASA/JPL)

of the Day from April 30, 2008, http://apod.nasa.gov/apod/ap980430.html. This one was actually taken from the Mars Reconnaissance Orbiter. If this happens, simply swap the colored lenses between your eyes.

Quite often, the stereo effect is exaggerated. You might argue that this is cheating, and therefore wicked, but a counter argument could be made to the effect that the exaggeration at least enables the observer to work out what he or she is looking at in a very unearthly place; and no one is being deceived about the exaggeration. The collection of 3-D images from Spirit and Opportunity given to the press is located at http://marsrovers.jpl.nasa.gov/gallery/press/opportunity/20050728a.html.

3-D images of this kind are sometimes called anaglyphs.

The rovers on the ground rely on artificial satellites in orbit around Mars to communicate with Earth. NASA has two orbiters at the time of writing: Mars Odyssey, http://mars.jpl.nasa.gov/odyssey, and Mars Reconnaissance Orbiter, http://mars.jpl. nasa.gov/mro. Each of these is performing many other missions besides acting as communications stations. The websites advertise that they are to investigate the climate and geology of the Red Planet. They also have those hardy perennials "to prepare for human exploration" and to "search for life on Mars." The Americans have been planning on putting someone on Mars "in the near future" for at least the last 40 years; and readers may by now have detected the author's skepticism about the existence of life there.

The European Mars Express has been in Martian orbit since the end of 2003. It tried to land a static science mission, *Beagle 2*. Unfortunately this was lost. At this distance it has not been possible to find out why. The orbiter, however, is still doing its thing. Its very accessible website is http://www.esa.int/Our_Activities/Space_Science/Mars_Express. There are some wonderful videos constructed from images obtained by this orbiter at http://www.esa.int/spaceinvideos/Missions/Mars_Express. You can always turn the music that accompanies some of these videos down if it irritates you.

India has an orbiter on its way, due to arrive in September 2014. The website is at http://www.isro.org/mars/home.aspx.

NASA sent a lander called Phoenix to Mars' northern polar icecap in 2008. It found water ice there. Of course, because it is so cold, and the pressure was so low, they were never going to find liquid water there. It also found evidence of carbonate rocks. These require water to form. This probe was placed in a hostile environment. It appears from the home page of the mission's website (http://phoenix.lpl.arizona.edu/gallery.php) that the solar panels were destroyed by frost. Nevertheless the mission succeeded in its objectives.

You can read the scientific papers about water in the Martian arctic at http://www.sciencemag.org/content/325/5936/58.full?ijkey=9ZTMoi3Mylweg&keytype=ref&siteid=sci, http://www.sciencemag.org/content/325/5936/64.full?ijkey=BVZRNinUWg62c&keytype=ref&siteid=sci and http://www.sciencemag.org/content/325/5936/61.full?ijkey=tmmqZ1fV6F9z.&keytype=ref&siteid=sci.

The latest rover to be sent to Mars is Curiosity (http://mars.jpl.nasa.gov/msl/). That this is truly a spacecraft for our age is shown by its ability to take selfies like the one shown (Fig. 11.12)

NASA's confidence in the ability of this spacecraft to last is indicated by the fact that it is designed to last. It is not that Spirit and Opportunity were not designed to last. They clearly must have been. But Curiosity is nuclear-powered. That investment would not have been made for a 90-day mission. The power comes from a radioisotope power system. This is not like a nuclear power station on Earth, or the nuclear propulsion systems in some warships, which generate steam from the heat of the nuclear reaction. Instead, the trickle of heat from the radioisotope is fed to thermocouples, wires made of dissimilar metals, which generate voltages. The radioisotope in question is plutonoium-238 (which cannot be used in weapons, before somebody worries).

The main advantage of this system over the solar power used for Spirit and Opportunity is that it will keep going at night and through dust storms. In principle it would also work in winter, although that is not much of an issue at Curiosity's latitude of 4.5°S.

Raw images from Curiosity can be found at http://mars.jpl.nasa.gov/msl/multimedia/raw. Processed images can be found at http://mars.jpl.nasa.gov/msl/multimedia/images. There are videos at http://mars.jpl.nasa.gov/msl/multimedia/videos and http://mars.jpl.nasa.gov/msl/multimedia/videoarchive.

Fig. 11.12 A self-portrait of the NASA Mars rover Curiosity (Image courtesy of NASA/ JPL-Caltech/Malin Space Science Systems. Source: http://photojournal.jpl.nasa.gov/catalog/ PIA16239)

From the videos you can learn little gems such as that the wheels have to have aluminum tires because Martian rocks are too sharp for rubber tires. (The thin atmosphere would also probably lead to rapid UV degradation of rubber.) Presumably the rocks are sharper because there is only thin air to erode them, unlike on Earth, where there is a thick atmosphere and lots of rain.

Jupiter

There have been five missions to Jupiter, plus a handful of others where Jupiter was used to provide gravity assist. The four missions are *Pioneer 10* (1973), *Pioneer 11* (1974), *Voyager 1* (1979), *Voyager 2* (1979) and *Galileo* (1995–2003). Dates in parentheses are the dates when these spacecraft flew by Jupiter, or, in the case of Galileo, orbited it.

Fig. 11.13 The four Galilean moons as imaged by the probe named for their discoverer (Image courtesy of NASA. Source: http://photojournal.jpl.nasa.gov/catalog/PIA01299)

Each of these missions sent back pictures better than any we had previously seen of the King of the Solar System. As has been discussed at length in this book, what we see from the ground has also improved massively since the mid-1970s. The best amateur astrophotographers can now do as well as *Pioneer 10* (see e.g., www. damianpeach.com).

The first four missions made it very obvious that Jupiter's Great Red Spot is a circulation storm, particularly the movie sequence shown at http://photojournal.jpl. nasa.gov/browse/PIA02855.gif. This movie was made from images taken by *Voyager 1*. One image was taken every Jovian day from January 6 to February 3, 1979. The author vividly remembers the stir these images caused in the physics common room when she was a graduate student.

Galileo completed 35 orbits in 9 years. Compare this to the period of Callisto, the most distant of the Galilean moons of Jupiter, which has a period of 16.7 days. From this we may deduce that the orbit was not especially close. In fact it was highly elliptical (i.e., oval) to enable the probe to travel widely in Jupiter's magnetosphere. The closest flyby – past the innermost moon Io – was left until last because of fears that the intense radiation might damage the probe. In fact it did give rise to the need for some remote computer maintenance.

The Galileo legacy website is at http://solarsystem.nasa.gov/galileo/index.cfm. The principal discoveries are listed at http://solarsystem.nasa.gov/galileo/discovery. cfm. The top ten science images appear at http://solarsystem.nasa.gov/galileo/gallery/top10science.cfm.

The other spectacular achievement of this mission was to detect the aftermath of the impact of Comet Shoemaker-Levy-9. The resultant images are shown at https://solarsystem.nasa.gov/galileo/gallery/comet-sl9.cfm.

The detailed images of the four Galilean satellites that we now take for granted were sent to us by the Galileo probe (Fig. 11.13). What would Galileo have thought about this probe named after him? What would the clerics who imprisoned him have thought, had they known how history would judge Galileo?

Saturn

This planet was explored later than Jupiter, in many cases by the same spacecraft, since they would usually hitch a ride from Jupiter's gravity. In particular, though, the Cassini/Huygens probe, which is still in orbit around Saturn at the time of this writing, arrived in mid-2004, after the end of the Galileo mission to Jupiter. Its photographic capability was therefore the technology of 1997, when it was launched.

As with Jupiter, the photos from early probes were state-of-the-art for their time. The Voyager pictures were a lot better than the Pioneer ones. The pictures of the main planet, however, are not quite as good as those of amateur astronomer Damian Peach. Compare http://www.damianpeach.com/sat13.htm with http://www.ciclops.org/view/3163/Saturn-taken-from-Voyager-2?js=1. This is not altogether a fair comparison. Some of the close-up photos from Voyager probes are better than either of these examples. For example, *Voyager 2* discovered spoke-like features in Saturn's rings (Fig. 11.14).

The pictures of Saturn's moons from the Voyager missions were of a different order from those made by amateurs. Figure 11.15 shows an example.

The Cassini/Huygens mission, which continues at the time of this writing, has been a tremendous success. Whereas the Galileo mission to Jupiter was greatly handicapped by a failed antenna [163], no such mishaps have befallen Cassini/Huygens.

Fig. 11.14 Spoke-like features in Saturn's rings discovered by Voyager 2 (Image courtesy of NASA. Source: http://photojournal.jpl.nasa.gov/catalog/PIA02275)

Fig. 11.15 Saturn's moon Mimas imaged by Voyager 1 in 1980 (Image courtesy of NASA. Source: http://photojournal.jpl.nasa.gov/catalog/PIA01968)

Early in 2005, the European Space Agency's Huygens probe made a soft landing onto Titan, having entered an atmosphere thicker than that on Earth. The technical skill required to do this a billion miles away beggars belief.

A movie showing the landing event is available at http://www.youtube.com/watch?v=HtYDPj6eFLc. If Beethoven's background music is not to your taste, you can always mute the video. There is no spoken commentary for you to miss. The video is so awesome that it needs no augmentation. The last sequence, from on the surface, seems to have been simulated from images such as the one at http://photojournal.jpl.nasa.gov/catalog/PIA07232. It would be unfair to say that 'simulated' means 'faked.' The surface features are quite recognizable from the still image.

NASA seems to have grasped that amateurs have a significant contribution to make to the analysis of the huge number of images sent back by Cassini. http://saturn.jpl.nasa.gov/photos/amateurimages shows how you can find raw images. It is then up to you how you process them to bring out the best in them. At the time of this writing, this web page cites three amateur astronomers: Italian Elisabetta Bonora, Maksim Kakitsev and Gordon Ugarkovic.

Maksim Kakitsev writes that he creates color images from raw monochrome images in the visible, infrared or ultraviolet region, using *PaintShop Pro* or *Photoshop.*

The world may or may not be yours to explore, but Saturn certainly is. Perhaps someone will be inspired to make enough images out of the data to write a book in the same series as this one.

There are other projects you can try. The author once asked Professor Carl Murray, an orbital dynamicist working on the Cassini mission, how amateurs could contribute. His answer was instant: discover moons from the images. His website is at http://www.maths.qmul.ac.uk/~carl. You would need to be capable of understanding how an orbit is calculated, but the opportunity is there.

The Ice Giants, Uranus and Neptune

The Voyager missions' data predates the Internet. Since these missions were launched in 1977, the cameras they carried were of an older generation than those we now take for granted.

Of the two Voyager missions, only *Voyager 2* went to Uranus and Neptune. It flew by Uranus in January 1986 and by Neptune in August 1989.

Although there is a *Voyager 2* website, at http://voyager.jpl.nasa.gov, it is not the greatest website for looking at interesting photos. There is a lot of information about the spacecraft itself, and a lot about the passage of *Voyager 2* through the heliosheath, the outer region of the heliosphere.

A better source of the photographs taken is Wikimedia Commons. There are 74 photographs of Uranus and its moons at http://commons.wikimedia.org/wiki/Category:Photos_of_Uranus_system_by_Voyager_2 and 77 photographs of Neptune and its moons at http://commons.wikimedia.org/wiki/Category:Photos_of_Neptune_system_by_Voyager_2.

If you wish to dig more deeply into the archives, you need to contact the Coordinated Request and User Support Office at the National Space Science Data Center, contact details at http://nssdc.gsfc.nasa.gov/about/about_cruso.html.

Asteroids

The first visit to an asteroid was the NEAR probe from NASA, which was launched in 1996 and landed on 433 Eros (the 433rd asteroid to be discovered) in 2001, after completing 230 orbits of the asteroid. Eros' orbit crosses that of Mars. In principle it might one day hit Earth [164]. If it did, it would make a mess: a not dissimilar one to that made by the asteroid that killed the dinosaurs. NEAR's landing on Eros, however, went better than planned. The spacecraft unexpectedly survived, and continued to transmit data for a couple of weeks. The website for the mission is at http://near.jhuapl.edu. Eros is shaped a little like a boomerang (well, all right, it is if you use a lot of imagination), and it is about 20 miles from end to end. The images of this asteroid are stored in date order at http://near.jhuapl.edu/iod/archive.html. There are also a few images of an en-route flyby of Asteroid 253 Mathilde.

In 2003 the Japan Aerospace Exploration Agency sent a mission to Asteroid 25143, called Itokawa. It brought samples back to Earth, arriving in 2010. The mission has a YouTube channel at http://www.youtube.com/user/jaxachannel?feature=watch. Quite a lot of the material on this channel is in English.

Fig. 11.16 A mosaic view of Vesta taken by the new Dawn spacecraft as it was leaving for Ceres (Image courtesy of NASA. Source: http://dawn.jpl.nasa.gov/multimedia/images/pia15678_full.jpg)

New Dawn was launched by NASA in 2007, and reached and orbited 4 Vesta, the fourth of the asteroids to be discovered, from July 2011 to September 2012, and then left for Ceres. It is due to get there in February 2015.

The mission's website is at http://dawn.jpl.nasa.gov. As is often the case with NASA websites, there are activities for children as well as adults.

Figure 11.16 shows a view of Vesta taken by the New Dawn space probe. The asteroid is not round. The mountain shown at the bottom is on the edge of the massive crater Rheasilvia. It is twice as high as Mount Everest.

Comets

At the time of this writing, four countries or groups of countries have sent nineteen missions to comets: the United States, the former Soviet Union, the European Space Agency and Japan. Most have been at least partially successful.

The Soviet Vega mission killed two birds with two stones. It put soft landers and balloons onto and into Venus, then hitched a gravitational slingshot ride to Comet Halley. There were two identical spacecraft, with instruments from Hungary, Czechoslovakia and France. Launched in 1984, they flew past Halley in March 1986. The images they obtained of the nucleus can be found at http://arc.iki.rssi.ru/IPL/vega.html.

The Japanese Sakigake probe only took non-imaging instruments to Halley. Its twin Suisei took UV images of the hydrogen corona. Suisei is Japanese for 'comet.'

Europe's Giotto made the closest approach to Halley's Comet on March 13, 1986. It flew by the nucleus of the comet, missing it by 370 miles. The imaging results were published in *Nature* [165]. There is a lovely picture of the comet's nucleus at http://sci.esa.int/science-e-media/img/3c/HalleyNucleus.jpg. The principal scientific findings are given at http://sci.esa.int/giotto/31878-halley.

NASA's Stardust, lunched in 1999, collected samples from comet Wild 2 in January 2004 and returned them to Earth in 2006. Its website is http://stardust.jpl.nasa.gov/home/index.html. This website gives a popular account of the scientific discoveries.

The NASA Probe Deep Impact went to Tempel 1 and dropped an impactor onto it. There is some spectacular footage of this impact. Chemical analysis of the ejecta was performed, and amateur astronomers were asked to track the comet and report brightness changes. This probe then went on to comet Hartley/2. You can read all about this on http://www.nasa.gov/mission_pages/deepimpact/main/index.html#.Uva6rYXkc-g. The impact occurred in July 2005. The ingredients of the comet are entertainingly presented at http://www.nasa.gov/mission_pages/deepimpact/media/spitzer-di-090705.html#.Uva8PIXkc-h.

Stardust, just mentioned, later flew on and photographed the artificial crater on Tempel 1. It found that the crater had partially re-filled, possibly with the original ejecta, possibly with new material that oozed out of the comet.

Kuiper Belt Objects

The New Horizons space probe is due to fly by Pluto on July 14, 2015. There is already a website for the probe: http://pluto.jhuapl.edu. On the way it got gravity assists from Mars, Jupiter, Saturn and Uranus. It holds the record for the fastest man-made object – over 36,000 miles per hour. If suitable targets can be identified, it will explore other Kuiper Belt objects after flying past Pluto.

As it passed Jupiter, this probe took pictures of the very volcanic plume that the Hubble Space Telescope photographed in February 2007 (Fig. 11.2). These images presage well for what New Horizons will do in the Kuiper Belt (Fig. 11.17).

Higher Hanging Fruit

For the determined armchair explorer with a lot of patience, there are archives of NASA's space probe data. There is no hiding from the fact that you need to be a scientist with the relevant specialty to decode the data from scientific instruments, but there are photographic archives. The place to look is the NASA Planetary Data System.

Recent funding cuts have not helped the maintenance of this online database. It provides a service to professional scientists, not to the general public. There is nothing to stop amateur astronomers having a good poke around, but you must expect to be communicated with as if you were a trained scientist.

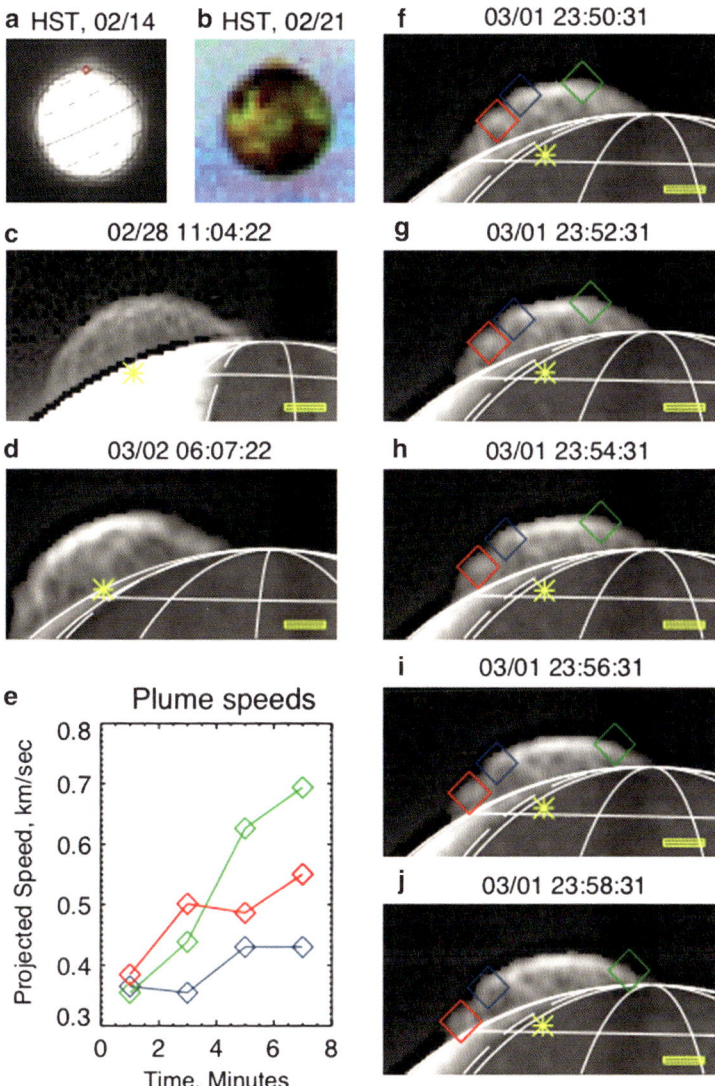

Fig. 11.17 The volcanic plume shown in Fig. 11.2, photographed by the Hubble Space Telescope, is shown in the top left, and compared to what New Horizons saw a little later, on March 1, 2007. The point that local telescopes see better than the Earth-orbiting Hubble is amply made (Source: http://pluto.jhuapl.edu/gallery/sciencePhotos/image.php?page=1&gallery_id=2&image_id=60. Credit: NASA/Johns Hopkins University Applied Physics Laboratory/Southwest Research Institute)

There are links to other sites such as the Planetary Photojournal, http://photojournal.jpl.nasa.gov. You can spend a lot of time surfing through this.

If you go much further, however, the searching quickly becomes a lot more difficult. A good starting point is the Planetary Image Atlas, http://pds-imaging.jpl.nasa.gov/search/search.html#QuickSearch. From this website the author quickly found 190,202 mono images from the mast cam on the Mars Science Laboratory, also known as Curiosity.

You may also need these two references:

Planetary Data System Standards Reference: http://pds.jpl.nasa.gov/documents/sr/StdRef_20090227_v3.8.pdf.

Planetary Science Data Dictionary Document: http://pds.nasa.gov/documents/psdd/PSDDmain_1r71.pdf

There are enough images there to keep you busy through rainy nights, even in those parts of the world where, if you stand still, moss grows on your north side.

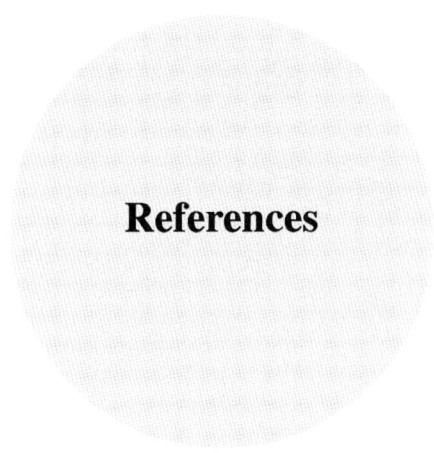

References

1. M.E. Bakich, *The Cambridge Planetary Handbook* (Cambridge University Press, Cambridge, 2000). ISBN 0521623803
2. http://nssdc.gsfc.nasa.gov/planetary/factsheet/sunfact.html
3. http://en.wikipedia.org/wiki/Mercury_%28planet%29
4. http://en.wikipedia.org/wiki/Venus
5. http://en.wikipedia.org/wiki/Moon
6. http://en.wikipedia.org/wiki/Mars
7. http://en.wikipedia.org/wiki/Asteroid_belt, http://en.wikipedia.org/wiki/Ceres_%28dwarf_planet%29
8. http://en.wikipedia.org/wiki/Jupiter
9. http://en.wikipedia.org/wiki/Saturn
10. http://en.wikipedia.org/wiki/Uranus
11. http://en.wikipedia.org/wiki/Neptune
12. A. Stern, J.E. Colwell, Collisional erosion in the primordial Edgeworth-Kuiper Belt and the generation of the 30–50 AU Kuiper Gap. Astrophys. J. **490**(2), 879–882 (1997)
13. A. Delsanti, D. Jewitt, *The Solar System Beyond the Planets* (Institute for Astronomy, University of Hawaii) Springer Praxis Books 2006, UK, pp 267–293
14. J. Wilkinson, *Probing the New Solar System* (CSIRO Publishing, Collingwood, 2008). ISBN 0643095756
15. S. Mitton, *The Cambridge Encyclopaedia of Astronomy* (Jonathan Cape, London, 1977)
16. http://en.wikipedia.org/wiki/Extrasolar_planet
17. G. de Santillana, *The Crime of Galileo,* New Edition (Chicago University Press, Chicago, 1978). ISBN 0226734811
18. K. Ferguson, *Measuring the Universe: The Historical Quest to Quantify Space* (Headline Book Publishing, London, 1999). ISBN 0747221324
19. A. Koestler, *The Sleepwalkers,* New Edition (Penguin, 1989). ISBN 0140192468
20. P. Moore, *Eighty Not Out* (Contender Books, London, 2003). ISBN 1843570483
21. http://neuronresearch.net/vision/files/adaptation.htm#a2
22. http://en.wikipedia.org/wiki/Planet

23. J.D. Clark, *Measure Solar System Objects and their Movements for Yourself!* (Springer, New York, 2009). ISBN 978-0-387-89560-4
24. http://amazing-space.stsci.edu/resources/explorations//groundup/lesson/scopes/rosse/index.php
25. J.L. Dobson, *How and Why to Make a User-Friendly Sidewalk Telescope*, 1st edn. (Everything in the Universe, Oakland, 1991). ISBN 0913399639
26. R. Wielen, H. Jahreiß, C. Dettbarn, H. Lenhardt, H. Schwan, Polaris: astrometric orbit, position, and proper motion. Astron. Astrophys. **2**(21), 1–13 (2000)
27. K.F. Riley, M.P. Hobson, S.J. Bence, *Mathematical Methods for Physics and Engineering. A Comprehensive Guide*, 3rd edn. (Cambridge University Press, Cambridge, 2006)
28. T.A. Littlefield, N. Thorley, *Atomic and Nuclear Physics: An Introduction in S. I. Units*, 2nd edn. (Van Nostrand Reinhold, London, 1968)
29. S. Gasiorowicz, *Quantum Physics*, 3rd edn. (Wiley, New York, 2003)
30. H. Muirhead, *The Special Theory of Relativity* (Macmillan, London, 1973)
31. P. Strange, *Relativistic Quantum Mechanics: With Applications in Condensed Matter and Atomic Physics* (Cambridge University Press, Cambridge, 1998)
32. W.E. Burcham, *Nuclear Physics: An Introduction* (Longman, London, 1963)
33. S. Perlmutter et al., Astrophys. J. **517**, 565 (1999)
34. A.G. Riess et al., Astrophys. J. **607**, 665 (2004)
35. W.K.H. Panofsky, M.N. Phillips, *Classical Electricity and Magnetism* (Addison-Wesley, Reading, 1962)
36. Christman, J, The Weak Interaction, http://physnet2.pa.msu.edu/home/modules/pdf_modules/m281.pdf
37. J.E. Greivenkamp, *Field Guide to Geometrical Optics* (SPIE Press, Bellingham, 2004). This book covers most of the optical concepts mentioned in this chapter.
38. J.D. Clark, Work Cited (2009)
39. I. Asimov, *Asimov on Astronomy* (Bonanza Books, New York, 1979)
40. M.A. Covington, *Digital SLR Astrophotography* (Cambridge University Press, Cambridge, 2007)
41. R. Berry, J. Burnell, *The Handbook of Astronomical Image Processing*, 2nd edn. (Willmann-Bell, Richmond, 2005)
42. R. Reeves, *Introduction to Webcam Astrophotography: Imaging the Universe with the Amazing, Affordable Webcam* (Willmann-Bell, Richmond, 2006)
43. E. Hubble, A relation between distance and radial velocity among extra-galactic nebulae, in *Proceedings of the National Academy of Sciences of the United States of America*, vol. 15(3), 15 Mar 1929, pp. 168–173
44. S. Hawking, *A Brief History of Time* (Bantam Dell, London, 1988). ISBN 9780553109535
45. D.N. Schramm, *The Big Bang and Other Explosions in Nuclear and Particle Astrophysics* (World Publishing Co., Singapore, 1994). ISBN 9810220243
46. http://en.wikipedia.org/wiki/Tests_of_general_relativity
47. S. Weinberg, *The First Three Minutes* (Fontana, London, 1977)
48. R.J. Bouwens et al., Nature **469**, 504–507 (2011)
49. R.J. Tayler, *The Stars: Their Structure and Evolution* (Cambridge University Press, Cambridge, 1994). ISBN 0521458854
50. E.M. Burbidge, G.R. Burbidge, W.A. Fowler, F. Hoyle, Synthesis of the elements in stars. Rev. Mod. Phys. **29**(4), 547 (1957)
51. C.D. Murray, S.F. Dermott, *Solar System Dynamics* (Cambridge University Press, Cambridge, 1999), pp. 42–45, Section 2.6
52. R.M. Canup, Forming a Moon with an Earth-like composition via a giant impact. Science **338**, 1052–1055 (2012)
53. See for example J.H. Jones, *Tests of the Giant Impact Hypothesis*, Lunar and Planetary Science. Origin of the Earth and Moon Conference (Monterey, 1998); A.E. Saal, et al., Volatile content of lunar volcanic glasses and the presence of water in the Moon's interior. Nature 454(7201), 192–195 (2008, July 10); and hang, Junjun; Nicolas Dauphas, Andrew M. Davis, Ingo Leya, Alexei Fedkin (25 March 2012). *The proto-Earth as a significant source of lunar material*. Nature Geoscience 5, 251–255

54. E.R.D. Scott, (December 3, 2001). *Oxygen Isotopes Give Clues to the Formation of Planets, Moons, and Asteroids*. Planetary Science Research Discoveries (PSRD). Bibcode 2001psrd. reptE..55S. Retrieved 2010-03-19

55. http://en.wikipedia.org/wiki/Orbit_of_the_Moon

56. E.C.T. Chao, E.M. Shoemaker, B.M. Madsen, First natural occurrence of coesite. Science, New Series, **132**(3421), 220–222 (1960)

57. B.A. Ivanov, A.T. Basilevsky, On the problem of central mounds/peaks formation in impact craters: observations and simulation. Lunar Planet. Sci. X, 607–609

58. H.J. Melosh, B.A. Ivanov, Impact crater collapse. Annu. Rev. Earth Planet. Sci. **27**, 385–415 (1999)

59. W.K. Hartmann, G.P. Kuiper, Concentric circles surrounding lunar basins. Commun. Lunar Planet. Lab. **1**(12), 51 (1962)

60. B.A. Ivanov, A.T. Basilevsky, On the problem of central mounds/peaks formation in impact craters: observations and simulation. Lunar Planet. Sci. X, 607–609 (1979)

61. http://www.nasa.gov/centers/ames/multimedia/images/2005/comets2a.html

62. B.A. Ivanov, V.N. Kostuchenko, Impact crater formation: dry friction and fluidization: influence on the scaling and modification, in *Lunar Planet. Sci. Conf. 29th, Abstr. 1654,* Lunar Planet. Inst., Houston

63. H.J. Melosh, B.A. Ivanov, Impact crater collapse. Annu. Rev. Earth Planet. Sci. **27**, 385–415 (1999)

64. G.R. Osinski, E. Pierazzo, *Impact Cratering: Processes & Products* (Wiley-Blackwell, Chichester, 2012)

65. I. De Pater, J.J. Lissauer, *Planetary Sciences* (Cambridge University Press, Cambridge, 2010), pp. 265–7. ISBN 9780521853712

66. M.A. Wieczorek et al., The constitution and structure of the lunar interior. Rev. Mineral. Geochem. **60**, 221–364 (2006)

67. F. Horz, R. Grieve, G.H. Heiken, P. Spudis, A. Binder, in *Lunar Surface processes, in Lunar Sourcebook*, ed. by G.H. Haiken, D.T. Vaniman, B.M. French (Cambridge University Press, Cambridge, 1991), pp. 61–120. ISBN 9780521334440

68. H.J. Melosh, *Planetary Surface Processes* (Cambridge University Press, Cambridge, 2011), p. 240. ISBN 9780521514187

69. www.damianpeach.com

70. J. Franklin, The renaissance myth. Quadrant **26**, 51–60 (1982)

71. http://en.wikipedia.org/wiki/Hubble_Space_Telescope

72. P.Y. Dely (ed.), *The Design and Construction of Large Optical Telescopes* (Springer, New York, 2003)

73. http://www.nasa.gov/mission_pages/messenger/timeline/index.html

74. http://solarsystem.nasa.gov/missions/profile.cfm?MCode=Cassini

75. M.M. Woolfson, *The Formation of the Solar System: Theories Old and New* (Imperial College Press, London, 2007), Chapter 38

76. H.S. Liu, J.A. O'Keefe, Theory of rotation for the planet mercury. Science **150**(3704), 1717 (1965)

77. M.E. Bakich, Work Cited, (2000), p. 30

78. Calculated using the software package *Cartes du Ciel*

79. P. Moore, *80 Not Out* (Contender Books, London, 2003)

80. J.S. Lewis, *Physics and Chemistry of the Solar System*, 2nd edn. (Academic, Burlington, MA, 2004)

81. J.D. Clark, *Measure Solar System Objects and their Movements for Yourself!* Patrick Moore's Practical Astronomy Series (Springer, New York, 2009)

82. A. Koestler, *The Sleepwalkers* (Grosset & Dunlap, New York, 1963)

83. A.R. Wallace, *Is Mars Habitable? A Critical Examination of Professor Percival Lowell's Book "Mars and Its Canals", with an Alternative Explanation, by Alfred Russel Wallace, F.R.S., etc.* (Macmillan and co., London, 1907)

84. F.B. Salisbury, Martian biology. Science **136**(3510), 17–26 (1962)
85. http://www.plospathogens.org/article/info:doi/10.1371/journal.ppat.0030055
86. http://uanews.org/node/20276
87. D. Tabor, *Gases, Liquids and Solids: And Other States of Matter*, 3rd edn. (Cambridge University Press, Cambridge, 1991)
88. http://nssdc.gsfc.nasa.gov/planetary/factsheet/marsfact.html
89. M.T. Lemmon et al., Atmospheric imaging results from the Mars exploration rovers: spirit and opportunity. Science **306**, 1753–1756 (2004)
90. M.E. Bakich, Work Cited, (2000), pp. 63–80
91. P. Moore, G. Hunt, *The Atlas of the Solar System* (Crescent Books, New York, 1990), p. 267
92. M.E. Bakich, Work Cited, information dispersed through the book
93. M.E. Bakich, Work Cited, (2000), p. 203
94. L.T. Elkins-Tanton, *Jupiter and Saturn* (Chelsea House, New York, 2006). ISBN 0-8160-5196-8
95. S.T. Weir, A.C. Mitchell, W.J. Nellis, Metallization of fluid molecular hydrogen at 140 GPa (1.4 Mbar). Phys. Rev. Lett. **76**(11), 1860 (1996)
96. F. Bagenal, T.E. Dowling, W.B. McKinnon, *Core mass: Jupiter: The Planet, Satellites and Magnetosphere* (Cambridge University Press, Cambridge, 2007). ISBN-10: 0521035457
97. T. Guillot, D.J. Stevenson, W.B. Hubbard, D. Saumon, The interior of Jupiter, in *Jupiter: The Planet, Satellites and Magnetosphere*, ed. by F. Bagenal, T.E. Dowling, W.B. McKinnon. Cambridge Planetary Science (Cambridge University Press, Cambridge, 2007), pp. 35–59
98. R. Helled, Constraining Saturn's core properties by a measurement of Its moment of inertia—implications to the Cassini Solstice Mission. Astrophys. J. Lett. **735**(1), article id. L16 (2011)
99. Shu Lin Li, C.B. Agnor, D.N.C. Lin, Embryo impacts and gas giant mergers. I. Dichotomy of Jupiter and Saturn's core mass. Astrophys. J. **720**, 1161–1173 (2010)
100. W.F. Denning, Jupiter, early history of the great red spot on. Mon. Not. R. Astron. Soc. **59**, 574–584 (1899)
101. Calculated with the software package *Cartes du Ciel*
102. C. Huyghens, *Systema Saturnium*, published a few years earlier as an anagram (1659)
103. Cassini's observation of the gap in Saturn's ring, now called Cassini's Division. Journal des Sçavans, 4 January 1677
104. J.C. Maxwell, Abstract of Professor Maxwell's paper on the stability of Saturn's rings. Mon Not R Astron Soc **19**, 297–304 (1859)
105. G.P. Kuiper, Titan: a satellite with an atmosphere. Astrophys. J. **100**, 378 (1944)
106. H.B. Niemann et al., The abundances of constituents of Titan's atmosphere from the GCMS instrument on the Huyghens probe. Nature **438**(7069), 779–784 (2005)
107. A. Coustenis, F.W. Taylor, *Titan: Exploring an Earthlike World*, 2nd edn. (World Scientific Publishing, Singapore, 2008)
108. http://www.nasa.gov/centers/ames/news/releases/2006/06_57AR.html
109. J.D. Clark, Work Cited, (2009)
110. R.J. Tayler, *The Sun as a Star* (Cambridge University Press, Cambridge, 1996)
111. A. Weiss, W. Hillebrandt, H.-C. Thomas, H. Ritter, *Cox and Giuli's Principles of Stellar Structure*, 2nd edn. (Cambridge Scientific Publishers, Cambridge, 2004)
112. D. Pitts, L.E. Sissom, *Schaum's Outline of Heat Transfer* (McGraw-Hill, New York, 1998)
113. S.V. Patankar, *Numerical Heat Transfer and Fluid Flow* (Taylor & Francis, Boca Raton, 1980)
114. H.K. Versteeg, W. Malalasekera, *An Introduction to Computational Fluid Dynamics, the Finite Volume Method* (Longman, Harlow, 1995)
115. H. Schwabe, Solar observations during 1843. Astron Nachr **20**(495), 233–236 (1843)
116. I. Newton, *Opticks*, Royal Society. An available edition with commentary by Nicholas Humez is (Octavo ed.) (Octavo, Palo Alto, 1704). ISBN 1-891788-04-3
117. Source: http://en.wikipedia.org/wiki/History_of_spectroscopy. Image is labelled as public domain

118. E.G. Loewen, E. Popov, *Diffraction Gratings and Applications* (Marcel Dekker, Inc., New York, 1997). ISBN 0824799232

119. D. Rittenhouse, An optical problem posed by F. Hopkinson and solved. J. Am. Phil. Soc. **201**, 202–206 (1786)

120. T. Young, On the theory of light and colours. Phil. Trans. **II**, 399–408 (1803)

121. J. Fraunhofer, Kurtzer Bericht von der Resultaten neurer Versuche über die Gesetze des Lichtes, und die Theorie derselbem. Gilberts Ann. Phys. **74**, 337–378 (1823)

122. H.J.J. Braddick, *Vibrations, Waves and Diffraction* (McGraw-Hill, London, 1966), pp. 175–213, Chapter 8

123. F. Mandl, *Statistical Physics* (Wiley, London, 1971), pp. 250–264, Chapter 10. ISBN 0471566586

124. E. Rutherford, Philos. Mag. **21**, 669–688 (1911)

125. S. Gasiorowicz, *Quantum Physics* (Wiley, New York, 2003). ISBN 0471057002

126. A.P. French, *Special Relativity* (CRC Press, Boca Raton, 1968). ISBN 0748764224

127. T.A. Littlefired, N. Thorley, *Atomic & Molecular Physics*, 2nd edn. (Van Nostrand Reinhold, London, 1968), pp. 92–133, Chapter 7. ISBN 0442048246

128. P. Zeeman, The effect of magnetisation on the nature of light emitted by a substance. Nature **55**(1424), 347 (1897)

129. G.E. Hale, On the probable existence of a magnetic field in sun-spots. Astrophys. J. **28**, 315 (1908)

130. G.E. Hale, F. Ellerman, S.B. Nicholson, A.H. Joy, Astrophys. J. **49**, 153–178 (1919)

131. G.E. Hale, F. Ellerman, S.B. Nicholson, A.H. Joy, The magnetic polarity of sun-spots. Astrophys. J. **49**, 153 (1919)

132. http://www.skyandtelescope.com/resources/proamcollab/3307266.html?showAll=y

133. A.R. Choudhuri, *The Physics of Fluids and Plasmas: An Introduction for Astrophysicists* (Cambridge University Press, Cambridge, 1998). ISBN 9780521555432

134. A.R. Choudhuri, An elementary introduction to solar dynamo theory, in *Kodai School on Solar Physics* (AIP Conference Proceedings 919), ed. S.S. Hasan, D. Benerjee (New York), pp. 49–73

135. http://solarscience.msfc.nasa.gov/predict.shtml

136. A.G. Kosovichev, Advances in global and local helioseismology: An introductory review, in *The Pulsations of the Sun and the Stars*, ed. by J.-P. Rozelot, C. Neiner. Lecture Notes in Physics, vol. 832 (Springer, Heidelberg, 2011), pp. 3–84. doi:10.1007/978-3-642-19928-8_1, Chapter 1

137. R.J. Tayler, *The Sun as a Star* (Cambridge University Press, Cambridge, 1997). ISBN 052146837X

138. C.D. Murray, S.F. Dermott, *Solar System Dynamics* (Cambridge University Press, Cambridge, 1999), pp. 456–466, Section 9.8

139. J. Baer, S.R. Chesley, Astrometric masses of 21 asteroids, and an integrated asteroid ephemeris. Celest. Mech. Dyn. Astron. **100**, 27–42 (2008)

140. S. Mitton (ed.), *The Cambridge Encyclopaedia of Astronomy* (Jonathan Cape, London, 1977)

141. http://dawn.jpl.nasa.gov

142. M.M. Woolfson, Work Cited, Chapter 36 (2007)

143. C.D. Murray, S.F. Dermott, Work Cited (1999)

144. C.D. Murray, S.F. Dermott, Work Cited, section 1.5 (1999)

145. http://www.moonphases.info/the-great-comet-scare-of-1910.html

146. L. Tolstoy, 1865–6, *War and Peace*, English edition (Wordsworth Classics, London, 2001)

147. I. Newton, *Mathematical Principles of Natural Philosophy (Philosophiae Naturalis Principia Mathematica)*, Translated by Andrew Motte (1846) First American Edition, New York, (1687), available online at http://rack1.ul.cs.cmu.edu/is/newton

148. D. Boulet, *Methods of Orbit Determination for the Microcomputer* (Willmann-Bell, Richmond, 1991)

149. M.M. Woolfson, Work Cited, section 39.2, (2007), pp. 280–283

150. D.H. Levy, *Shoemaker by Levy: The Man Who Made an Impact* (Princeton University Press, Princeton, 2000)
151. http://www.nasa.gov/mission_pages/soho/comet-2000.html
152. http://cometchasing.skyhound.com
153. http://www.ast.cam.ac.uk/~jds
154. N. James, G. North, *Observing Comets*. Patrick Moore's Practical Astronomy Series (Springer, London, 2002)
155. http://www.theguardian.com/politics/blog/2009/apr/24/gordon-brown-angry
156. Source: http://en.wikipedia.org/wiki/File:Qr-3.png. Author: Autopilot, licensed under the Creative Commons Attribution-Share Alike 3.0 Unported license
157. http://www.spacetelescope.org/about/general/instruments/costar
158. R.J. Tayler, *The Sun as a Star* (Cambridge University Press, Cambridge, 1997), p. 31. ISBN 052146837X
159. W.J. Markiewicz et al., Nature **450**, 633–636 (2007); G. Piccioni et al., Nature **450**, 637–640 (2007); P. Drossart et al., Nature **450**, 641–645 (2007); J.-L. Bertaux et al., Nature **450**, 646–649 (2007); S. Barabash et al., Nature **450**, 650–653 (2007); T.L. Zhang et al., Nature **450**, 654–656 (2007); M. Pätzold et al., Nature **450**, 657–660 (2007); C.T. Russell, T.L. Zhang, M. Delva, W. Magnes, R.J. Strangeway, H.Y. Wei, Nature **450**, 661–662 (2007)
160. N. Slack, C. Chambers, C. Harland, A. Harrison, R. Johnston, *Operations Management*. Part 4: Improvement, 2nd edn. (Prentice-Hall, Harlow, 1998), pp. 677–794. ISBN 0273624962
161. http://en.wikipedia.org/wiki/Gusev_%28Mars_crater%29
162. http://en.wikipedia.org/wiki/Meridiani_Planum
163. http://jpl.nasa.gov/news/press_kits/gllarpk.pdf
164. P. Michel, P. Farinella, C. Froeschlé, The orbital evolution of the asteroid Eros and implications for collision with the Earth. Nature **380**(6576), 689–691 (1996)
165. H.U. Keller, C. Arpigny, C. Barbieri, R.M. Bonnet, S. Cazes, M. Coradini, C.B. Cosmovici, W.A. Delamere et al., First Halley multicolour camera imaging results from Giotto. Nature **321**(6067s), 320–326 (1986)

Index

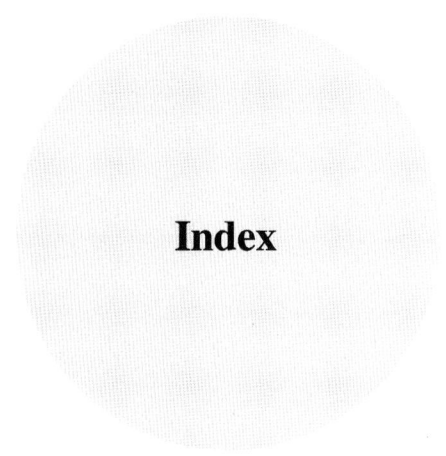

© Springer Science+Business Media New York 2015
J. Clark, *Viewing and Imaging the Solar System*, The Patrick Moore
Practical Astronomy Series, DOI 10.1007/978-1-4614-5179-2